vademecum

T0261911

Parasites of Medical Importance

Anthony J. Nappi, Ph.D.
Department of Biology
Loyola University
Chicago, Illinois, U.S.A.

Emily Vass, Ed.D.
Department of Biology
Loyola University
Chicago, Illinois, U.S.A.

CRC Press
Taylor & Francis Group
Boca Raton London New York

CRC Press is an imprint of the
Taylor & Francis Group, an **informa** business

VADEMECUM
Parasites of Medical Importance

First published 2002 by Landes Bioscience

Published 2018 by CRC Press
Taylor & Francis Group
6000 Broken Sound Parkway NW, Suite 300
Boca Raton, FL 33487-2742

ISBN 13: 978-1-57059-679-7 (pbk)

**Visit the Taylor & Francis Web site at
http://www.taylorandfrancis.com**

**and the CRC Press Web site at
http://www.crcpress.com**

Library of Congress Cataloging-in-Publication Data

CIP information applied for but not received at time of publishing.

Dedication

With the hope that it will live up to her high standards and expectations, this book is dedicated with affection to Emily, co-author, colleague and friend. She was a young scholar who always strived to learn more than the basics. She enjoyed her work, and it was a joy to work with her. Á toute á l'heure.

Emily Vass
11 February 1967 - 19 April 2002

Contents

Preface

Virtually every organism serves as the host for a complement of parasites. Parasitism is so common that it is rare to find classes of animals without members that have adopted a parasitic mode of living. Evidence gained from various archeological studies indicates that parasitic diseases existed in prehistoric human populations. Since there is no evidence to suggest that our long and intimate association with parasites will ever end, it seems reasonable to propose that the study of human parasites warrants some consideration. However, the study of parasites is a very challenging endeavor. Host-parasite associations involve complex biochemical, physiological, behavioral and ecological adaptations that very likely have co-evolved independently and on many different occasions. These complex and intimate interactions are continually evolving as counterstrategies in both host and parasite populations, thus limiting our ability to adequately study the factors that influence immune competency, parasite virulence, adaptability, epidemiological diversity, and drug resistance. However, the most important challenge facing parasitologists derives not from technical or experimental difficulties, but from the fact that most of the parasitic diseases that have a major impact on humans are largely associated with the rural poor in tropical, developing countries, which typically attract little interest from strictly commercial enterprises and other agencies that fund research.

Today, the extent of human suffering due to parasites is incalculable and intolerable. The physiological, pathological and economic problems caused by parasites are global concerns, and it is imperative that health professionals have some understanding of the complex interactions between humans and their parasites. Inexplicably, many medical schools fail to offer a curriculum that contains a formal course in parasitic diseases, or, in some cases, even to provide a single lecture on the topic. It is our belief that the collaborative efforts of parasitologists and medical professionals are urgently needed to improve efforts to treat parasitic infections. *Parasites of Medical Importance* is designed primarily for health professions and students interested in pursuing careers that will address the growing threat current and emerging parasitic diseases pose to the global population. In preparing this textbook we assumed that it would be a first exposure to the study of parasites for those who have had little or no formal instruction in parasitic diseases. Thus, emphasis has been placed on parasite life cycles and host pathology, with limited discussions of parasite morphology, taxonomy, and pharmacological treatments.

The authors assume full responsibility for omissions or any mistakes that appear in the book, and will correct such issues in subsequent editions.

Acknowledgements

We wish to thank Dr. Pietro Carmello of the Carlo Denegri Foundation, Torino, Italy, for granting permission to use several of the illustrations that are maintained by the Foundation. We wish to acknowledge the Centers for Disease Control, Division of Parasitic Diseases, Atlanta, Georgia and the Bayer AG Company, Leverkusen, Germany, for providing several illustrations. A special thanks to the following colleagues who provided us with original photographs: Harvey Blankespoor, Robert Kuntz and Dianora Niccolini. A portion of the effort spent on finishing the textbook was made possible because of research support from The National Institute of Health (GM 59774), The National Science Foundation (IBN 0095421) and Research Services at Loyola University Chicago.

CHAPTER 1

Interspecific Interactions

The term symbiosis, which means the "living together" of two species, was first used in 1879 by the botanist Heinrich Anton de Bary to describe the relation between certain species of fungi and algae living together to form lichens. Based primarily on the type of dependency that exists between the interacting species, three types of symbiosis are distinguished; commensalism, mutualism, and parasitism.

Commensalism is a type of symbiotic association which is beneficial to one species and at least without any detectable adverse effect on the other species. The basis for such a relation may be food, substrate, space, or shelter. The commensal is usually the smaller of the two species and may be attached to the exterior of the host (ectocommensal), or live within the body of the host (endocommensal). Examples include certain tropical commensal fishes, which are protected from predation by living among the tentacles of certain sea anemones, and the pilot and remora fishes, which associate with sharks, sea turtles, or other species of fish usually feeding on "leftovers". If the association is only a passive transport of the commensal by the host, the relationship is referred to as phoresy. Phoresy is essentially an accidental association with no metabolic dependency or interaction between the two individuals.

Mutualism is an association of two species that are metabolically dependent on each other. Examples of mutualism include flagellates living in the gut of wood roaches and termites, lichens, and the cultivation of fungi by various species of insect. Parasitism is an association of heterospecific organisms during which the parasite, usually the smaller of the two species, derives its nutrient requirements directly from (and at the expense of) the host. In some heterospecific interactions it is difficult to determine the nature of the symbiotic association because variations exist in the degree of intimacy, pathogenicity, and permanency of the association. Parasites living within the body of their hosts are termed endoparasites, while those attached to the outer surface of the body are called ectoparasites. The term infection is commonly used when discussing endoparasites, and infestation when reference is made to ectoparasites. Parasitoses is the infection or infestation of a host with animal parasites.

Parasitism may be the only option for an organism, or it may be an alternative way of life. If an organism is completely dependent on its host during all or a part of its life cycle and cannot exist in any other way, the parasite is known as an obligatory parasite. A facultative parasite is an organism that does not depend on the parasitic way of life at any stage during its development, but may become parasitic if provided the opportunity to do so. Multiple parasitism occurs when a host is infected (or infested) by two or more species of parasites, whereas superparasitism is the infection of a host by more individuals of a single species of parasite than the host can support. The host may be so severely injured by the heavy infection that, if it does not succumb, it provides such an inadequate environment for the parasites that they fail to develop completely and eventually die. The term superinfection is used when an infected

Parasites of Medical Importance, by Anthony J. Nappi and Emily Vass.
©2002 Landes Bioscience.

host is reinfected with the same species of parasite. If two or more hosts are involved in the life cycle of a parasite, the host in which the parasite reaches sexual maturity is termed the final or definitive host. Hosts associated with larval or juvenile stages of a parasite are referred to as intermediate hosts. A biological vector is a host that is not only required for the development of the parasite, but also for transferring the parasite to another host. A transfer or paratenic host is one that is not absolutely necessary for the completion of the parasite's life cycle, but serves as a temporary refuge and/or mechanical vector for transfer to an obligatory host. Hosts that serve as a direct source from which other animals can be infected are known as reservoir hosts. The term zoonoses refers to those diseases transmittable to humans from other animals.

Specificity in Host-Parasite Relations

Specificity refers to the mutual adaptations that restrict parasites to their hosts. A high degree of host specificity indicates that the parasite is unable to survive in association with any other species. The human pinworm, *Enterobius vermicularis*, and the beef tapeworm, *Taenia saginata*, are examples of parasites that are very host specific. Some of the factors that prevent a parasite from infecting an organism other than the host species include host immunity, seasonal, behavioral, or geographic barriers, or the absence of specific metabolites, intermediate hosts or biological vectors that are required for parasite development.

Host specificity may be a function of physiological, ecological, and/or behavioral adaptations. The conditions determining the degree of host specificity often are markedly different for the various developmental stages of a parasite that uses different hosts to complete its life cycle. Parasites with two or more intermediate hosts are less specific than those with one intermediate host. Also, parasites that infect the host by penetrating the skin tend to be more host-specific than those that are ingested by the host. Even within a single host the physiologic demands of the different stages of a parasite may be so different that there is site specificity (blood, liver, etc.) at different times during development. Generally, a parasite that has a high degree of host specificity requires a specific site within its host in which to develop, while a parasite that is not host specific lives in various host tissues. The beef tapeworm, which is specific for humans, can live only in the small intestine. On the other hand, the roundworm *Trichinella spiralis*, which infects various warm-blooded animals, can live in different host tissues. Unfortunately, very little is known of the factors that determine the localization of parasites within their hosts. The host tissue-specific sites occupied by parasites represent specific niches, and complex behavioral and physiological adaptations regulate the precise migratory routes followed by the parasites in locating these sites for their development.

Modes of Infection

The life cycles of parasites are characteristically complex, with many specific requirements for development and survival. Parasites with a direct life cycle develop in or on the body of only a single, definitive host. These parasites generally have a free-living stage away from the host, and adaptations for the successful transfer of this stage often include a protective covering (i.e., cuticle, thickened cell wall or cyst) and/or locomotor structures that propel the parasite in the environment. Parasites with indirect life cycles contend not only with environmental problems, but also with different biotic requirements of the definitive and intermediate host(s) that often belong to different phyla. Natural transfer of the infective stage(s) of a parasite may be accomplished by ingestion of contaminated food or water, inhalation, inoculative transmission during feeding of an infected host (e.g., trypanosomes, malaria), or by the active penetration of the host body by the parasite (cercariae of blood flukes, hookworm larvae). There may be transplacental transmission (*Toxoplasma gondii*), as well as via sexual intercourse (*Trichomonas vaginalis, Treponema pallidum*). Parasites may escape from their hosts by actively penetrating their tissues and by passage through the digestive, urinary, respiratory, or reproductive systems.

Clinical Effects of Animal Parasitoses

The adverse effects a parasite has on a host depend on numerous factors including host age, health, immune competence, nutritional state, site of attack, number of parasites, and the interaction of various environmental factors. In some host-parasite interactions there may be no pathological symptoms of infection (asymptomatic), while in others the parasites may produce clinically demonstrable effects. Unfortunately, the pathologies caused by animal parasites are not always diagnostically specific, and these may be mistaken for a variety of bacterial, fungal, or viral infections. Hence, positive identification of the parasite is always essential for effective treatment. Some examples of parasite-induced injuries include:

1. **Tissue Damage.** Injuries to tissues may occur during and/or after penetration of the host. Examples include scabic mites, fly maggots, ticks, penetration of hookworms, and mosquito punctures. The migration through the host body of eggs and larval stages of various helminths produce tissue lesions. Also, lytic necroses may result from enzymes released by tissue-inhibiting parasites.

2. **Stimulation of Host Cellular and Tissue Reactions.** Parasites and/or their metabolites may induce various inflammatory and immune responses by the host. Blood disorders may include eosinophilia, erythropoiesis, anemia, polymorphonuclear leukocytosis, and leukopenia. The salivary and venomous secretions of insects and other arthropods may provoke systemic responses such as allergic and neurological reactions in addition to localized skin inflammation at the site of the wound. Tissue abnormalities may involve fibrosis, granulomatous growths, metastasizing sarcomas, and carcinomas. In various cell types there may be evidence of hyperplasia (accelerated rate of mitosis), hypertrophy (increase in size), and metaplasia (abnormal cellular transformations). The production of antibodies (immunoglobulins) and the mobilization of phagocytic cells may in part characterize the immune response to various parasites.

3. **Mechanical Interference.** The invasion of numerous parasites into the body may cause partial or total obstruction of the digestive system and associated organs, circulatory system, and the lymphatic system. Considerable necroses of these organs are also manifested in heavy infections.

4. **Nutritional Disturbances.** Parasites acquire nutrients by consuming a

portion of the food ingested by the host, and/or by feeding directly on host cells, tissues, or body fluids. Host metabolism may be severely disturbed by the presence of parasites, and symptoms of a chronic nature such as gradual loss of weight and progressive weakness may develop. Parasite-induced pathogenicity may be manifested in response to inadequate host nutrition.

5. **Secondary Infections.** Many parasites produce ulcerations and wounds as they enter the host. These areas subsequently become sites for infection by microbial pathogens. Secondary microbial infections may be more serious than those caused by the parasites.

Prevalence of Parasitic Diseases

The incidence of human infection with parasites is staggering. These global problems are magnified because numerous other parasites ravage livestock, reduce agricultural productivity, and contribute greatly to serious nutritional deficiencies in underdevelped countries. The extent of suffering due to parasites is incalculable. Various sources estimate that approximately one billion persons are infected with the roundworm *Ascaris lumbricoides*, 700 million suffer from filariasis, 270 million have schistosomiasis, and 20 million suffer from trypanosomiasis. Each year, between 300 and 500 million people contract malaria, of whom between 1.5 and 2.7 million die. The World Health Organization estimates that one-fifth of the world population is under threat from this disease. Malaria and other mosquito-borne diseases (e.g., dengue fever, yellow fever, meningitis, filariasis) cause a death every 30 seconds. Either our present medical technology is inadequate to cope with these parasitic infections or our priorities need to be altered. Tropical medicine does not occupy a position in the mainstream of biomedical and clinical research because parasitic diseases generate little journalistic attention, and because the billions of people who suffer from tropical disorders are mostly poor, illiterate, and are seldom heard from. These problems represent global concerns with billions of individuals at risk.

CHAPTER 2

Major Groups of Parasites of Humans

I. Protozoa. Unicellular eukaryotic organisms
II. Helminths. Parasitic worms
Phylum 1. Platyhelminthes: Flatworms
Flukes
Tapeworms
Phylum 2. Nemathelminthes: Nematodes or unsegmented roundworms
III. Arthropods. Animals with chitinous exoskeleton and jointed appendages
Insects, spiders
Mites, Ticks
Scorpions, Lice
Fleas

Major Groups of Parasitic Protozoa

Phylum 1. Sarcomastigophora

These protozoans possess monomorphic nuclei, and flagella, pseudopodia, or both types of locomotor structures. They reproduce asexually by binary and multiple fission, typically without spore formation, and sexually by fusion of isogametes or anisogametes.

Subphylum Mastigophora (Flagellates)

Important species infecting humans include *Giardia lamblia, Leishmania tropica, L. braziliensis, L. donovani, Trichomonas vaginalis, Trypanosoma rhodesiense, T. gambiense* and *T. cruzi.*

Subphylum Sarcodina (Amoebas)

Important parasitic amoebas include *Acanthamoeba, Endolimax nana, Entamoeba histolytica, Entamoeba polecki, Entamoeba gingivalis, E. coli, E. hartmanni, Hartmannella, Iodamoeba butschlii* and *Naegleria fowleri.*

Phylum 2. Ciliophora

Very few parasitic forms are present in this subphylum. Simple cilia or compound ciliary organelles are present at some stage in their development. In most members the nuclei are of two types. Asexual reproduction is by binary fission; sexuality involves conjugation, autogamy, or cytogamy. The single important species is *Balantidium coli.*

Phylum 3. Apicomplexa

All members of this group are intracellular parasites without locomotor organelles. A complex system of organelles is present in the apical end at some stage. One or

Parasites of Medical Importance, by Anthony J. Nappi and Emily Vass.
©2002 Landes Bioscience.

more micropores is generally present. The organisms reproduce sexually and/or asexually, and in some members a cyst stage is present.

Class Sporozoa.

Subclass 1. Coccidia
Important species include *Isospora belli*, *Plasmodium malariae*, *P. vivax*, *P. falciparum*, *P. ovale*, *Sarcocystis lindemanni* and *Toxoplasma gondii*.

Subclass 2. Piroplasmia
Two important species are *Babesia bigemina* and *Theileria parva*.

Other Apicomplexa
The taxonomic status of some species remains questionable: One important member is *Pneumocystis carinii*.

Protozoan Reproduction and Life Cycles

Protozoans are typically microscopic and unicellular, and possess one or more nuclei and other organelles comparable to the cells of metazoan organisms. Protozoan parasites cause more suffering, debilitation and death than any other group of pathogenic organisms. The success of this group is attributed in large measure to their high reproductive potential.

Parasitic protozoans reproduce by asexual and/or sexual methods. Asexual methods include schizogony, or multiple asexual fission, and budding. In schizogony the nucleus and certain other organelles undergo repeated divisions before cytokinesis; the nuclei become surrounded by small amounts of cytoplasm, and cell membranes form around them while they are within the mother cell which becomes known as a schizont. The daughter cells, termed merozoites, are liberated when the cell membrane of the schizont ruptures. If multiple asexual fission follows the union of gametes, the process is termed sporogony. Budding involves mitosis with unequal cytokinesis.

Sexual reproduction in parasitic protozoa involves reductional division in meiosis resulting in a change from diploidy to haploidy, with a subsequent restoration of diploidy by the union of gametes (syngamy) derived from two parents (amphimictic), or from a single parent (automictic). When only haploid nuclei unite, the process is called conjugation. Gametes may be similar in appearance (isogametes) or dissimilar (anisogametes). Marked dimorphism is frequently seen in anisogametes. The larger gamete (female) is termed macrogamete, the smaller, generally more active gamete (male) is the microgamete. Fusion of gametes produces a zygote. Frequently, the zygote is a resting stage that overwinters or forms spores that enable survival during transfer to different hosts.

Another mechanism of transfer between hosts is encystment. In some parasitic forms, the normal feeding or vegetative stage (trophozoite) cannot infect new hosts because it cannot survive the transfer. Such protozoans secrete a resistant covering around themselves and enter a resting stage or cyst. In addition to protection against unfavorable conditions, cysts may serve for cellular reorganization and nuclear division. Possible adverse conditions within the host favoring cyst formation include desiccation, deficiency of essential host metabolites, changes in pH, temperature, or tonicity. In the group Sporozoa, the cyst is termed an oocyst. Within the oocyst sporogony and cytokinesis occur to produce infective stages termed sporozoites. The oocyst may serve as a developmental capsule for the sporozoites within the host, or it may be the resistant stage that is transmitted to new hosts.

Parasitic Flagellates

Flagellates constitute the largest group of parasitic protozoa. Typically, the body of a flagellate is elongate and slender with a single flagellum. However, some species are spheroid in shape, possess more than a single flagellum, or lack flagella entirely. The flagellum, which arises from a basal body or kinetoplast, may originate near, and extend freely from, the anterior end, or it may run along the free margin as an undulating membrane attached to the side of the organism. In some cases the flagellum passes through the entire body, extending beyond the anterior end of the organism as a free structure. Based on characteristics of the flagellum, as many as four developmental stages occur in the life cycles of flagellates that are transmitted by insects to humans (Fig. 1). The amastigote is a spheroid form, devoid of an external flagellum. A small internal flagellum extends only slightly beyond the flagellar pocket. This stage is found in the life cycles of the three species of *Leishmania* parasitizing humans. The promastigote is an elongate form with a kinetoplast located in front of the nucleus (antenuclear), near the anterior end of the organism. A short flagellum arises near the kinetoplast and emerges from the anterior end of the organism. The epimastigote is an elongate form. The kinetoplast is close to the nucleus (juxtanuclear) with a flagellum arising near it and emerging from the side of the organism to run along a short undulating membrane. The trypomastigote is an elongate form with a post nuclear kinetoplast and is the definitive stage of the genus *Trypanosoma*. The flagellum emerges from the side of the organism to run along a long undulating membrane, which is directed anteriorly. Two additional morphological stages of flagellates are known, the choanomastigote, and opisthomastigote. The choanomastigote form is slightly ovoid and has an antenuclear kinetoplast. The flagellum emerges from a wide funnel or collar-like reservoir at the anterior end of the body. The opisthomastigote is an elongate form with a post nuclear kinetoplast. The flagellum passes through the organism and emerges from its anterior end. There is no undulating membrane present. Neither the choanomastigote nor the opisthomastigote form occurs in the life cycle of any flagellate parasite of humans.

The flagellate parasites of humans generally reproduce asexually by longitudinal binary fission. Based on their location within their hosts, two medically important groups are recognized; flagellates of the blood and connective tissues (hemoflagellates), and flagellates of digestive or reproductive systems. Hemoflagellates require a blood-sucking arthropod as a biological vector, while flagellates of the digestive and reproductive passages do not.

Hemoflagellates: *Trypanosoma*

The flagellates that parasitize the blood and tissue of vertebrates belong to the family Trypanosomidae. There are two important genera of *Trypanosoma*tids, *Leishmania* and *Trypanosoma* (Table 1).

Most of the Trypanosomatids that parasitize terrestrial vertebrates require a blood-sucking arthropod as a biological vector. Two mechanisms of transmission of

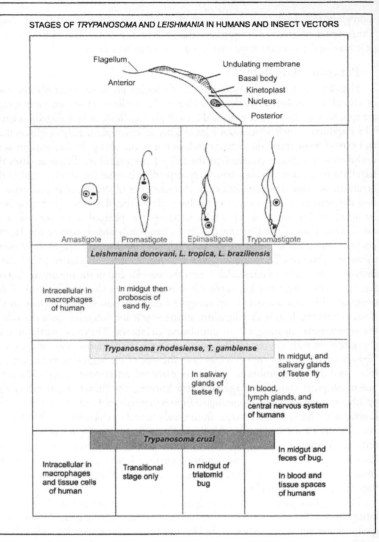

STAGES OF *TRYPANOSOMA* AND *LEISHMANIA* IN HUMANS AND INSECT VECTORS

Figure 1. Morphological stages of *Trypanosoma* and *Leishmania* found in humans and insect vectors. Modified from Beaver PC, Jung RC. Animal Agents and Vectors of Human Disease. Philadelphia: Lea and Febiger, 1985.

hemoflagellates occur with blood-sucking arthropod vectors. In one, the parasites pass from the mouth parts of the blood-feeding vector directly into the definitive host. This inoculative transmission is referred to as infection from the anterior station. In the second method, the parasites are voided in the feces of the biological vector, and infection occurs when the parasites are inadvertently rubbed into the wound produced by the vector when it bites the definitive host. This mechanism of infection resulting from wound contamination is referred to as infection from the posterior station. The only trypanosome of vertebrates not transmitted by an animal vector is *T. equiperdum*. This flagellate, which causes dourine in horses and mules, is transmitted during coitus.

Genus Trypanosoma

Most species of the genus *Trypanosoma* are parasites of the blood, lymph and tissue fluids of vertebrates. In these hosts, they appear in the trypomastigote form and divide by longitudinal binary fission. One notable exception is *T. cruzi*, which has become adapted to intracellular life in the amastigote form and does not multiply in the trypomastigote form. There are two major human diseases caused by trypanosomes, sleeping sickness, a disease found in Africa, and Chagas' disease which occurs in Central and South America, and parts of the United States.

African Trypanosomiasis (Sleeping Sickness)

Based on their separate geographic distributions and generally different clinical manifestations, two forms of African sleeping sickness are distinguished; Gambian (chronic) or West African form caused by *T. gambiense*, and a more virulent Rhodesian (acute) or East African form caused by *T. rhodesiense*. These two species are morphologically indistinguishable from each other, and from a third species, *T. brucei*, which infects many domestic and natural game animals, but apparently does not parasitize humans. *Trypanosoma gambiense* and *T. rhodesiense* are transmitted to humans by both sexes of the tsetse fly *Glossina*. *Glossina palpalis* and *G. tachinoides* are the principal biological vectors of *T. gambiense*, while those of *T. rhodesiense* are *G. morsitans*, *G. pallidipes*, and to a lesser extent, *G. swynnertoni*.

The trypanosomes of humans typically are found in the blood, lymph, spleen, liver, and cerebrospinal fluid (Fig. 2). When the tsetse fly bites and takes a blood meal from an infected individual, the flagellates are taken into the midgut of the insect where development begins. The trypanosomes later migrate into the proventriculus, labial cavity, and then into the salivary glands where they develop to the infective or metacyclic stage. The complete life cycle in the insect requires 2 to 3 weeks. Human infection occurs during host feeding when an infected tsetse fly injects the parasites contained in the saliva into the skin. In the area of the inoculation the parasites initiate an interstitial inflammation that gradually subsides within a week. Occasionally ulcerations appear at the site of the puncture with the formation of an indurated, painful chancre, which slowly disappears. Within 1 to 2 weeks after infection, the parasites gain access to the circulatory system and cause a heavy parasitemia, chills, fever, headache, and occasionally nausea and vomiting. Congenital infection is also possible with the parasites passing through the placenta. Breast milk from infected individuals also may be a source of infective trypanosomes.

The Gambian form of sleeping sickness involves primarily lymphoid and nervous tissues. Marked lymphadenitis occurs with the painful enlargement of the posterior cervical lymph nodes (Winterbottom's sign). Slaves from Africa en route to the Caribbean exhibiting such enlarged cervical lymph nodes were routinely thrown overboard by slave traders. Some of the more pronounced clinical manifestations as the disease advances include edema, enlargement of the spleen and liver, anorexia,

Table 1. Major blood and tissue-dwelling flagellates of humans

Parasite	Epidemiology	Location in Host	Mode	Symptoms of Infection
Leishmania tropica	Mediterranean area, Asia, Africa, Central America	Skin	Bite of *Phlebotomus* (sandfly)	Skin lesion
Leishmania braziliensis (espundia)	Central and South America	Skin and mucocutaneous tissue	Bite of *Phlebotomus*	Skin lesions, enlarged liver and spleen, death
Leishmania donovani (kala-azar)	Mediterranean area, Asia, Africa, South America	Skin and somatic organs	Bite of *Phlebotomus*	Skin lesions, enlarged liver and spleen, death
Trypanosoma gambiense (Gambian sleeping sickness)	West Africa	Blood, lymph nodes, central nervous system	Bite of *Glossina* (tsetse fly)	Lymph-adenopathy (Winter-bottom's sign), meningo-encephalitis enlarged liver and spleen, lethargy, death
Trypanosoma rhodesense (Rhodesian sleeping sickness)	Eastern and Central Africa	Blood, lymph nodes, central nervous system	Bite of *Glossina* (tsetse fly)	Enlarged liver and spleen, Glomerulo-nephritis, meningo-enceph-alitis, death
Trypanosoma cruzi (Chagas' disease)	United States, Central and South America	Cardiac muscle, blood and other tissues	Reduviid bugs by posterior station, congenital, ingestion of infected mothers' milk	Muscle pain, lympha-denitis, meningoen-cephalitis, myocarditis, tachycardia, death (Romana's sign)

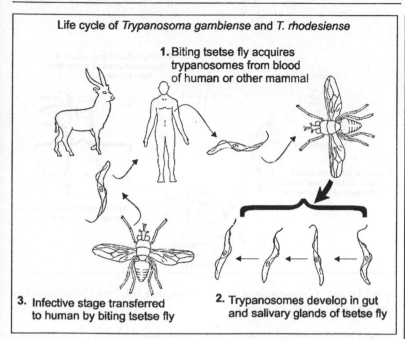

Life cycle of *Trypanosoma gambiense* and *T. rhodesiense*

1. Biting tsetse fly acquires trypanosomes from blood of human or other mammal

3. Infective stage transferred to human by biting tsetse fly

2. Trypanosomes develop in gut and salivary glands of tsetse fly

Figure 2. Life cycle of *Trypanosoma gambiense* and *T. rhodesiense*. Modified from Belding, D. L. 1958. Clinical Parasitology. Appelton-Century-Crofts, Inc., New York.

extreme weakness, rapid loss of weight, disturbed vision, meningoencephalitis, fibrillation of facial muscles, tremor of the tongue and hands, mental deterioration, paralysis, convulsions, and finally coma and death. The complete course of the disease may extend over several years. Rhodesian trypanosomiasis is a more rapid form of the disease than the Gambian form, usually resulting in death within a few months. Generally, there is little or no neurologic involvement associated with the disease, since rarely does the host live long enough for the parasites to attack the central nervous system. Domestic animals serve as reservoir hosts for both Gambian and Rhodesian trypanosomiasis. Native game animals are believed to serve as reservoirs for *T. rhodesiense*, but not for *T. gambiense*.

South American Trypanosomiasis (Chagas' Disease)

Trypanosoma cruzi is a parasite that lives in the blood and reticuloendothelial tissues of humans and many domestic and wild mammalian reservoir hosts, including dogs, cats, bats, raccoons, foxes, opossums, squirrels, monkeys and pigs. The geographic range of the parasite extends from southern parts of the United States through Mexico, Central and South America. Approximately 12 million persons are infected with *T. cruzi*. The principal vectors of *T. cruzi* are various reduviid bugs, including *Panstrongylus megistus*, *Triatoma infestans* and *Rhodnius prolixus*. The insects are notorious, nocturnal household pests, having a penchant for biting around the

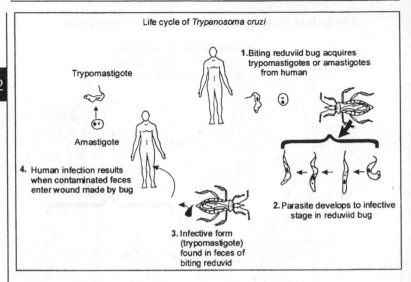

Life cycle of *Trypanosoma cruzi*

Trypomastigote

Amastigote

1. Biting reduviid bug acquires trypomastigotes or amastigotes from human

2. Parasite develops to infective stage in reduviid bug

3. Infective form (trypomastigote) found in feces of biting reduvid

4. Human infection results when contaminated feces enter wound made by bug

Figure 3. Life cycle of *Trypanosoma cruzi*. Modified from Belding, D. L. 1958. Clinical Parasitology. Appelton-Century-Crofts, Inc., New York.

eyes and lips of sleeping individuals. When feeding on the blood of infected vertebrates, the reduviids obtain the parasite either as free flagellates in the trypomastigote stage, or as intracellular amastigotes within host macrophages. With the blood meal the parasites pass first into the midgut of the insect where development transforms the flagellates into epimastigotes. The latter migrate to the hindgut where they are further transformed into infective or metacyclic trypomastigotes. The complete cycle in the insect requires about 2 weeks. Parasitized insects can retain an infection for several months (Fig. 3). While infected bugs feed, they defecate, voiding feces containing infective trypomastigotes. Human infection occurs when contaminated feces enters the skin through punctures made by the biting bugs, through skin abrasions, or through mucous membranes of the eye and mouth that are rubbed with contaminated fingers. Human infection may also occur through ingestion of the insect vector or its contaminated feces. The parasites also may be transmitted through the placenta and in breast milk. In endemic areas transmission may occur from infected donors during blood transfusions.

Entrance of the infective trypomastigotes initiates an acute local inflammation. During the early stages of infection, the trypomastigotes are abundant in the blood, but they do not undergo multiplication there. They eventually invade, and/or are engulfed by reticuloendothelial cells of the liver, spleen, and lymphatics, glial cells, and myocardial and skeletal muscles. Other tissues infected include nervous, gonadal, bone marrow and placenta. Within the various host cells, the trypanosomes rapidly transform into amastigotes that repeatedly multiply by binary fission producing numerous individuals. The parasites transform successively into promastigote, epimastigote, and trypomastigote stages, and are liberated when the destroyed host

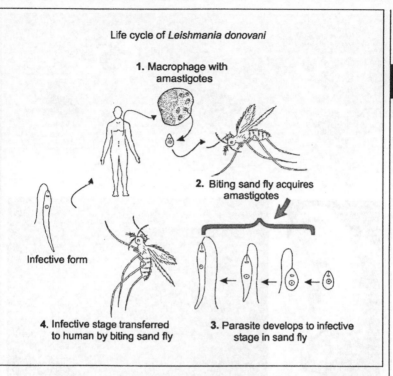

Life cycle of *Leishmania donovani*

1. Macrophage with amastigotes

2. Biting sand fly acquires amastigotes

Infective form

4. Infective stage transferred to human by biting sand fly

3. Parasite develops to infective stage in sand fly

Figure 4. Life cycle of *Leishmania donovani*. Modified from Belding, D. L. 1958. Clinical Parasitology. Appelton-Century-Crofts, Inc., New York.

cells rupture. The released trypomastigotes are infective to other host cells, as well as to the insect vectors. A generalized parasitemia accompanies the release of trypanosomes from host cells into the blood. Although almost every type of tissue is susceptible to invasion by *T. cruzi*, the flagellates demonstrate a preference for muscle and nerve tissues.

Chagas' disease appears in an acute stage primarily in young children and in a chronic form in adults. Frequently, early symptoms of the acute form appear as inflammatory swellings or nodules (chogomas) at the sites of the insect bite, unilateral edema of the eyelid and conjunctiva, and swelling of the pre-auricular lymph nodes (Romana's sign). Later, there is enlargement of the spleen, liver and lymphatic tissues, anemia, fever, and headache. Myocardial and neurological dysfunctions represent more severe manifestations of the chronic form of the disease. The heart becomes markedly enlarged and flabby. In endemic areas, the disease accounts for approximately 70% of the cardiac deaths in adults. Chagas' disease has been reported as the most important cause of myocarditis in the world. Additional manifestations include enlargement of the esophagus and colon, resulting in impaired peristalsis.

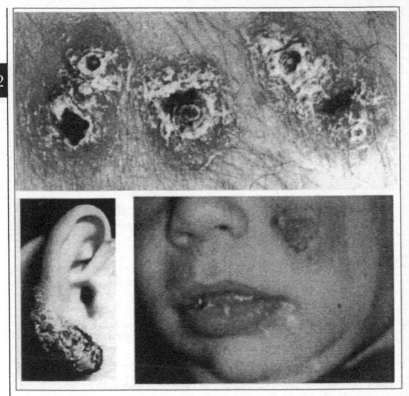

Figure 5. Examples of cutaneous Leishmaniasis caused by various species of *Leishmania*, including *L. tropica, L. mexicana* and *L. major*. Courtesy of Drs. Joseph El-On and Luis Weinrauch, Ben-Gurion University of Negev, Israel, and the Carlo Denegri Foundation, Torino, Italy.

Hemoflagellates: *Leishmania*

Genus Leishmania

Species of the Genus *Leishmania* occur in tropical and subtropical areas, where they are transmitted to humans and reservoir hosts (dogs, rodents) by female sand flies belonging to the genera *Phlebotomus* and *Lutzomyia*. The parasites occur in the amastigote stage within macrophages and reticuloendothelial cells of subcutaneous tissues, mucous membranes, liver, spleen and lymph nodes. Infected host cells rupture, liberating amastigotes that are engulfed by other phagocytes. When a sand fly sucks blood from an infected animal, amastigote forms are taken into the midgut where they transform into spindle-shaped promastigotes. The promastigotes multiply by binary fission, and migrate into the pharynx and buccal cavity from which they are injected into the skin of a vertebrate host when the fly again takes a blood meal. Mechanical transfer through the bite of stable flies (*Stomoxys calcitrans*) has been reported. Contact infection is possible when infected flies are crushed into the

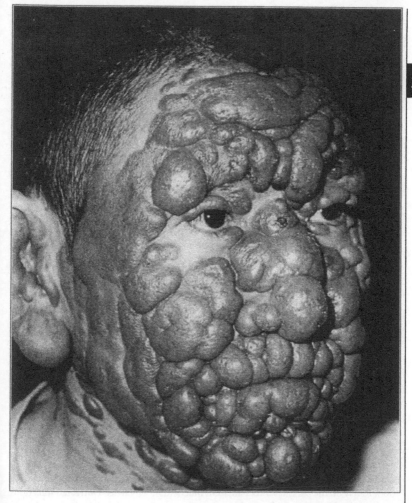

Figure 6. A severe dermal, post-visceral manifestation of "kala-azar" infection caused by *Leishmania donovani*. Photograph courtesy of Dr. Robert Kuntz.

skin or mucous membranes. Infection also may be possible by fecal contamination, since promastigotes have been found in the hindgut of some flies. After being introduced into the skin, the promastigotes are phagocytosed by macrophages, in which cells they undergo transformation to amastigotes, and begin to multiply. Heavily infected cells rupture, liberating amastigotes that are engulfed by other host macrophages, and parasite reproduction continues (Fig. 4).

The medically important species of *Leishmania* include *L. tropica*, *L. major*, *L. donovani*, *L. braziliensis* and *L. mexicana*. The parasites are morphologically indistinguishable and have virtually identical life cycles. They differ clinically and

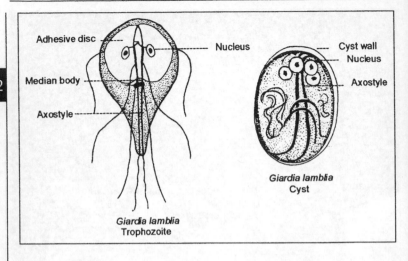

Figure 7. Diagrams of the trophozoite and cyst stages of *Giardia lamblia*.

serologically, but at times these characteristics overlap, thus species distinctions are not always clearly observed. *Leishmania tropica* and *L. major* are the etiological agents of cutaneous Leishmaniasis, also known as oriental sore, Jericho boil, Aleppo boil, or Delhi boil (Fig. 5). The disease occurs in Africa, the Middle East, southern Europe, Asia, India, Central and South America. The sand fly, *Phlebotomus papatasii*, is the important biological vector of cutaneous Leishmaniasis. After an extremely variable incubation period ranging from several weeks to three years, a small red sore or papule appears at the site of inoculation. Multiple sores may appear because of several infective bites or as a result of early contamination of other areas. The organisms may also disseminate within the host producing subcutaneous lesions of the face and appendages. Early papules gradually increase in size and become scaly. Ulceration occurs and spreads circularly. The lesion remains shallow, with a bed of granulation tissue, and surrounded by an area of red induration. The surrounding lymph nodes may become enlarged, especially if there is secondary bacterial infection. Rarely do the parasites infect adjacent mucocutaneous areas. Untreated infections gradually heal within several months to a year, leaving flattened and depigmented scars.

    Leishmania donovani causes visceral Leishmaniasis also known as dum-dum fever or kala-azar. The flagellates infect cells of the reticuloendothelial system throughout the body. Infections occur primarily in the spleen, liver, bone marrow, and visceral lymph nodes. *Leishmania donovani* occurs in the Mediterranean region, southern Russia, China, India, Bangladesh, Africa, and Central and South America. The parasites are transmitted by various species of *Phlebotomus,* including *P. argentipes, P. longipalpis* and *P. orientalis*. The incubation period varies from a few weeks to eighteen months. The parasites initially colonize the dermis and later enter the blood, lymphatics and then the viscera where they are engulfed by macrophages. Typically, the liver and spleen become greatly enlarged (hepatosplenomegaly). There is an increased production of macrophages, decreased erythropoiesis, and thrombocytope-

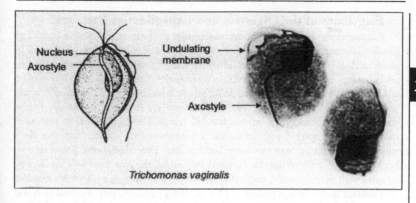

Figure 8. Diagram and stained preparations of *Trichomonas vaginalis*.

nia, which results in multiple hemorrhages. As the disease progresses there is a gradual loss of weight, the abdomen becomes swollen by the enlargement of the viscera. Other symptoms include edema, especially of the face, breathing difficulties, chills, fever, vomiting, and bleeding of the gums, lips, mucous membranes, and intestinal mucosa. In some individuals, there develops a post-kala-azar dermal leishmanoid condition, characterized in part by reddish, depigmented nodules in the skin (Fig. 6). The mortality in untreated cases may reach 95%. A fatal outcome is common in infected infants and young children. Death generally occurs within three years after infection.

Leishmania braziliensis is the etiological agent of a disease variously known as mucocutaneous Leishmaniasis, espundia or uta. The geographical range of the parasite is from Mexico to Argentina. The clinical manifestations of the disease, reservoir hosts, and species of sand flies involved in transmission, vary considerably from one location to another. Among several species of phlebotomine sand flies that serve as vectors are *Lutzomyia flaviscutellata*, *L. intermediua* and *L. tropidoi*. At the site of inoculation, a primary lesion, similar to oriental sore, occurs. This primary lesion heals within 6-15 months. Secondary lesions, characterized by epithelial hyperplasia, inflammation, and edema, may develop on the ear (chiclero ulcer), causing erosion of the earlobe cartilage. Secondary lesions may also occur in the mucous membranes of the mouth and nose (espundia), causing erosion of the lips, nasal septum, palatine tissues, pharynx, larynx, and trachea. The time of appearance of secondary lesions varies from before the primary lesions heal to many decades after infection. Death may result from secondary infections and/or respiratory complications.

Leishmania mexicana produces a disease with cutaneous, nasopharyngeal mucosal, and visceral manifestations. The cutaneous form of the disease is called "chiclero ulcer", and is common in individuals harvesting gum from chicle trees. The parasite is found in Texas, Mexico, and parts of Central America. Sand flies of the genus Lutzomyia are biological vectors. The disease is a zoonosis with rodents as the principal reservoir host.

Flagellates of the Digestive and Reproductive Passages

Giardia lamblia is the most common flagellate of the human digestive tract. The parasite is cosmopolitan, but the disease, giardiasis, is more commonly found in children than in adults, and in individuals residing in warm climates rather than in cold climates. In some areas in the United States the incidence of infection may be as high as 20% of the population. The pathogen has both trophozoite and cystic stages in its life cycle (Fig. 7). The trophozoites are confined essentially to the duodenum, but occasionally invade the bile ducts. The trophozoite is 'tear-drop shaped', with a convex dorsal surface and a concave ventral surface ('adhesive disc') which makes contact with the intestinal mucosa. The trophozoite possesses two nuclei, and four pairs of flagella. In heavy infections, the intestinal mucosa may be carpeted with the parasites. A single diarrhetic stool from a heavily infected individual may contain several billion parasites. The flagellates penetrate into mucosal cells and also interfere with the absorption of fat and fat soluble vitamins. Heavy infections may be characterized by extensive ulcerations of the intestinal mucosa. Biliary disease sometimes occurs when flagellates pass up the bile duct. The trophozoites multiply by longitudinal binary fission in the small intestine and eventually encyst. Mature cysts, which are tetranucleate, are found in stools. Infection of new hosts occurs when mature cysts are ingested with contaminated food or water. Following excystation in the duodenum, the tetranucleate parasite undergoes cytokinesis forming two binucleate daughter trophozoites, which then adhere to epithelial cells and feed. Symptoms of this highly contagious disease include diarrhea, abdominal pain, the passing of blood and mucus, hypoproteinemia with hypogammaglobulinemia, fat-soluble vitamin deficiencies, and the production of copious light-colored fatty stools.

Trichomonas vaginalis is a cosmopolitan parasite that resides in the male and female urogenital tracts. Transmission of the infective trophozoite stage is chiefly by sexual intercourse, and because of its potentially pathogenic nature, the disease is regarded as a serious venereal disease (Fig. 8). Transmission may occur from female to female through contaminated clothing or toilet facilities. The parasite has been found in newborn infants. In males, infection is frequently asymptomatic, but severe symptoms involve not only the urethra (urethritis) and bladder, but also the genital organs and glands, including the prostate. A discharge from the urethra containing the flagellates may occur. Although the vagina is most commonly infected (vaginitis), the trichomonads may spread to all parts of the urogenital tract of the female. A frothy, creamy discharge is frequently observed in infected females. The disease may be complicated by concurrent fungal, bacterial, or spirochetal infections.

Some Nonpathogenic Flagellates

Trichomonas tenax is a nonpathogenic species confined to the mouth, especially in pyorrheal pockets and tarter along the gumline, and in tonsillar crypts. Transmission of trophozoites may result from kissing or the use of common drinking or eating utensils. Other nonpathogenic intestinal flagellates include *Dientamoeba fragilis*, *Chilomastix mesnili*, *Retortamonas intestinalis*, *Enteromonas hominis* and *Pentatrichomonas hominis*. Frequently, the presence of these nonpathogenic forms is an indication of direct fecal contamination.

CHAPTER 3

Sarcodina

Amoebae belong to the subphylum Sarcodina, a group comprised mostly of free-living organisms living in a wide variety of habitats. Symbiotic amoebae commonly found in the intestinal tract of humans and domestic animals belong to the genera *Entamoeba, Endolimax* and *Iodamoeba* (Fig. 1). Species infecting extraintestinal sites include members of the genera *Naegleria, Hartmannella* and *Acanthamoeba*. The life cycle may involve just a motile, feeding trophozoite stage, or both trophozoite and cyst stages. In most members the trophozoite is amoeboid only, moving by means of pseudopodia. In some life cycles, both flagellate and amoeboid trophozoites are represented. Reproduction involves mitotic divisions of the nucleus followed by binary fission. Cysts may be ingested, or possibly even inhaled. Excystation generally occurs in the small intestine of the host.

Entamoeba histolytica is a pathogenic, tissue-invading amoeba. Perhaps 80-95% of the one-half billion human infections by this parasite are symptomatic. The highest incidence of the disease, termed amoebiasis, occurs in persons living in warm climates, in rural areas where poor sanitary conditions exist, and in crowded institutions such as prisons and asylums. As with most enteric diseases, amoebiasis is commonly associated with sewage contamination of drinking water. The virulence of *E. histolytica* varies considerably. Unfortunately, the factors which determine the virulence of the parasite are not completely understood. Dietary deficiencies appear to influence the incidence and severity of infection.

Infection occurs by the ingestion of mature cysts of the parasite in contaminated food or drinking water, and by hand-to-mouth contact (Fig. 2). Flies and cockroaches may mechanically transport the cysts and contaminate food and eating utensils. Following excystation in the ileum, the emerging trophozoites, occasionally referred to as metacysts, multiply by binary fission and adhere to the intestinal mucosa. They lodge in the crypts of the lower portions of the small intestine and in the large intestine, where some invade the mucosal epithelium by elaborating a proteolytic enzyme which lyse the cells. The parasites may invade deep into the wall of the intestine, feeding, eroding tissues, and forming ulcers. The majority of intestinal lesions occur in the cecal and sigmoid-rectal areas. Invading amoebae may enter mesenteric venules or lymphatics and be transported to the liver, lungs, brain, and other extraintestinal organs where they continue feeding, causing severe lesions and tissue necrosis. Some of the intestinal trophozoites do not invade the gut wall but instead form cysts (a process termed encystation) which are later passed outside the host. Immature cysts contain one or two nuclei; mature cysts contain four nuclei. Immature cysts are able to mature in the external environment and become infective. Trophozoites passed in the feces are unable to encyst outside the host and are not infective.

Based on the anatomic site of infection and on clinical manifestations, two major types of symptomatic amoebiases are recognized: (1) Intestinal or primary amoebiasis, including both dysenteric and non-dysenteric forms; and, (2) extraintestinal or

Parasites of Medical Importance, by Anthony J. Nappi and Emily Vass.
©2002 Landes Bioscience.

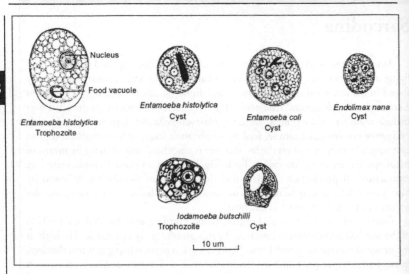

Figure 1. Drawings of representative amoebae.

secondary amoebiasis, including hepatic, pulmonary, cutaneous, cerebral, coronary and urogenital sites. Symptomatic intestinal infections of *E. histolytica* are termed amoebic colitis. Common, relatively mild symptoms include diarrhea, dysentery (blood and mucus in stools), abdominal discomfort, flatulence (gas in stomach or intestine), anorexia (loss of appetite), and loss of weight. The pathology of intestinal amoebiasis is destruction of the intestinal epithelium and invasion of the gut wall by the trophozoites. The initial invasion by the parasites into the intestinal mucosa may initiate an inflammatory response. The perforation of amoebic ulcers and resulting peritonitis characterize moderately severe infections. In the liver, single or multiple abscesses may occur, and the liver frequently becomes enlarged. Extensive invasion of the liver results in the destruction of parenchymal tissue by cytolysis. Hepatic infection may spread through the diaphragm to the lungs (amoebic pneumonitis) and respiratory passages where abscess formation frequently follows. Abscess formation in these areas is frequently accompanied by the infiltration of leukocytes and fibroblasts.

Naegleria fowleri causes a serious disease in humans termed primary amoebic meningoencephalitis. The parasite rapidly invades and multiplies in the brain and meninges causing extensive hemorrhage and tissue destruction. Infections, which are nearly all fatal, have been reported from New Zealand, Australia, East Africa, Europe, and America. The disease resembles bacterial meningitis. The first signs of infection are mild fever, headache, nausea, vomiting, and in some cases, a sore throat. Later, as the headache and fever persist and increase in severity, the patients become disoriented and usually comatose. Death usually occurs within seven days of the onset of the symptoms. The infective trophozoite stage is diphasic in that it exists not only as an amoeba, but under certain conditions, such as when transferred from a culture medium to distilled water, it develops two flagella to become a flagellate. In

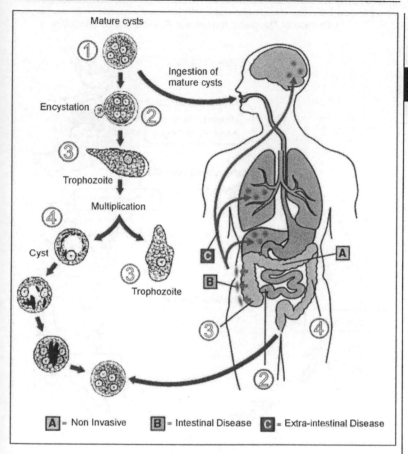

Figure 2. Life Cycle of *Entamoeba histolytica*. Following the ingestion of cysts (1), excystation occurs in the small intestine (2). The trophozoites (3) may remain in the intestinal lumen of individuals (A), who then become asymptomatic carriers that can disseminate the cysts, or the trophozoites invade the intestinal mucosa (B), enter the cisrculation and establish extraintestinal infections (C). Courtesy of the Centers for Disease Control, Division of Parasitic Disease, Atlanta, Georgia.

human tissues, the parasite is present only in the amoeboid phase. A resistant cyst stage is formed only in laboratory cultures. Intracerebral infection is generally acquired while the individual is swimming in fresh or brackish water, with the mode of entry into the brain and meninges through the olfactory mucosa and cribiform plate. Apparently, some infections are acquired by inhaling airborne trophozoites (Fig. 3).

Species of *Hartmannella* and *Acanthamoeba* also invade the central nervous system. Members of these two genera are similar morphologically to the amoeboid stage of *Naegleria*, but do not possess flagella at any stage in their developments. *Hartmannella* has been observed in the respiratory tract of humans. *Acanthamoeba*

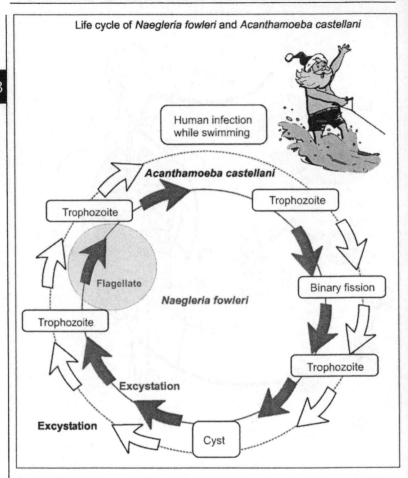

Life cycle of *Naegleria fowleri* and *Acanthamoeba castellani*

Figure 3. Comparison of the life cycles of *Naegleria fowleri* and *Acanthamoeba castellani*.

is believed to cause ulcerations of the cornea, resulting in blindness, and lesions and granulomatous formations of the skin, especially in immune compromised individuals (Fig. 4). Cysts of *Acanthamoeba* have been found in cases of meningoencephalitis.

Entamoeba gingivalis is a common commensal found in the mouth, especially in the gingival areas, in the tartar near the base of the teeth, and in the crypts of the tonsils. No cysts are formed by *E. gingivalis*; only the trophozoite stage has been described. Transmission is by droplet spray, intimate oral contact, or by the sharing of eating utensils. Although *E. gingivalis* is frequently associated with peridontitis, there is no evidence that the organism is pathogenic.

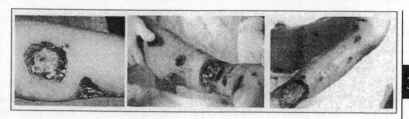

Figure 4. Cutaneous acanthamoeba infection in a 35 year-old individual with AIDS and previous episodes of *Pneumocystis carinii* pneumonia and toxoplasmic encephalitis. The lesions are initially nodular and reddish, but later become ulcerative, necrotic and more numerous. Courtesy of Drs. Antonio Macor, Ezio Nigra, Romina Ruffatto, Alberto Pisacane and M.L. Soranzo, at the Amedeo di Savoia Hospital, Torino, Italy and the Carlo Denegri Foundation, Torino, Italy.

Entamoeba coli, E. hartmanni, E. polecki, Iodamoeba butschlii and *Endolimax nana* are all regarded as nonpathogenic intestinal amoebae. Their life cycles are similar, with infection occurring via excystation in the intestine. Although generally harmless, the presence of the organisms indicates ingestion of contaminated food or water.

Ciliate Parasites

Balantidium coli is the only ciliate parasite of humans. It is cosmopolitan, inhabiting the large intestine, cecum, and terminal ileum, feeding on bacteria. The parasite may also invade the intestinal mucosa causing ulceration. Extraintestinal spread to the liver, lungs, and urogenital tract is rarely observed. Infection occurs by ingestion of cysts in contaminated food or drink. Less than one percent of the human population is infected with *B. coli*. Pigs represent the usual source of infection for humans. Symptoms of the disease, which is termed balantidiasis, may range from mild colitis and diarrhea to clinical manifestations resembling severe amoebic dysentery.

CHAPTER 4

Apicomplexa: Sporozoa and Piroplasmea

Introduction to Sporozoa

All members of the Class Sporozoa are parasites. The Class consists of two subclasses, Gregarina and Coccidia. The Gregarina are parasites of the intestinal tract, body cavity, and reproductive organs of invertebrates only. The coccidians live in the intestinal mucosa, liver, reticuloendothelial cells, blood cells, and other tissues of vertebrates and invertebrates. The latter group includes parasites causing serious coccidioses of domestic animals, and malarias of humans and other vertebrates. The major species parasitizing humans belong to the genera *Isospora, Toxoplasma, Cryptosporidium, Babesia, Pneumocystis* and *Plasmodium*.

Coccidians

In many coccidians, reproduction is both asexual, by either binary or multiple fission, and sexual, by either isogamous or anisogamous union. Spore formation frequently follows sexual reproduction. The infective stage is the sporozoite. Transmission from host to host is accomplished either by vectors, usually biting arthropods, which transmit the infective sporozoites directly, or by ingestion of highly resistant cysts (oocysts) with internal spores (sporocysts) containing the infective sporozoites.

The typical Coccidian life cycle pattern consists of three developmental phases: sporogony, schizogony, and gamogony (Fig. 1). Different morphological types result from each developmental phase; sporozoites from sporogony, merozoites from schizogony, and zygotes from gamogony. The only diploid stage in the life cycle is the zygote. Following meiosis, the zygote becomes a multinucleate sporont and undergoes multiple fission (sporogony). Depending on the species, the haploid cells formed are either free infective sporozoites or sporozoites enclosed within an oocyst or spore, from which they subsequently escape. Within the host, infective sporozoites become intra- or extracellular vegetative trophozoites. The trophozoites develop first into multinucleate schizonts, which then undergo multiple fission (schizogony) producing numerous daughter cells or merozoites. Merozoites either continue the infection of the host with repetition of asexual multiplication and invasion of other host cells, or they develop into sexual stages termed gamonts. Gamonts undergo multiple fission (gamogony) and give rise to gametes. The latter join in pairs in syngamy to produce diploid zygotes.

Variations in the basic life cycle pattern among different species result from the absence of any one phase. When sporogony is omitted, the zygotes develop directly into sporozoites, and when gamogony is omitted, merozoites develop into gametocytes directly. In monoxenous coccidians, all stages in the life cycle occur in a single host, with the oocyst eliminated from infected individuals. In heteroxenous forms,

Parasites of Medical Importance, by Anthony J. Nappi and Emily Vass.
©2002 Landes Bioscience.

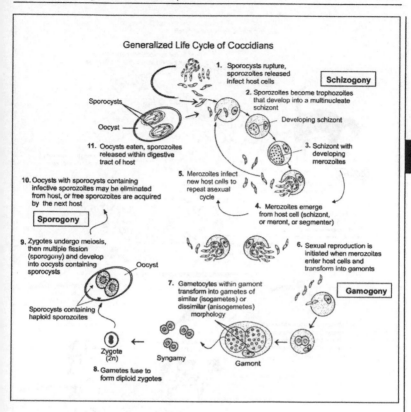

Figure 1. Developmental stages in the life cycle of coccidians.

sporogony occurs in an invertebrate host, and following the transfer of sporozoites by biting vectors, schizogony and a part of gamogony occur in the vertebrate host.

Genus Isospora

Isospora belli is a parasite of the intestinal mucosa of humans. Although widely distributed, *I. belli* is more prevalent in warm climates than in cool climates. The life cycle of this species is believed to be similar to those of *I. canis* in dogs and *I. felis* in cats. Infection occurs when mature or sporulated oocysts are ingested. The freed sporozoites invade the intestinal mucosa where all three developmental phases occur. Asexual stages terminate with the production of oocysts, which are eliminated in the feces. The oocysts of *Isospora* contain two sporocysts, and within each sporocyst there are four infective sporozoites.

Apparently, the majority of human infections are without adverse manifestations, but symptoms ranging from mild gastrointestinal discomfort, nausea, and anorexia, to severe dysentery or diarrhea have been reported. The stools are often loose and pale yellow, indicating, as in giardiasis, the inability of the patient to adequately absorb fat. Infections may last up to four months.

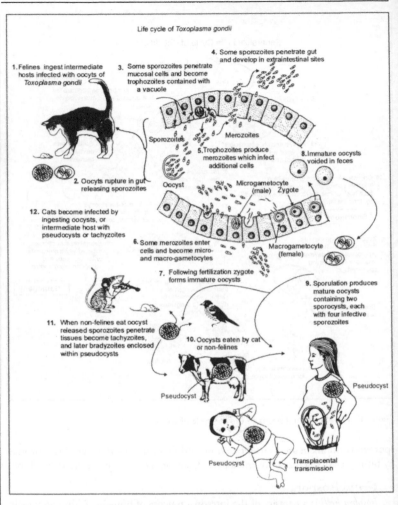

Figure 2. Life cycle of *Toxoplasma gondii*. Modified from Schmidt, G. D. and Roberts, L. S. 1996. Foundations of Parasitology. Wm. C. Brown Publishers.

Genus Toxoplasma

Toxoplasma gondii is a cosmopolitan parasite of a variety of mammals and birds. The domestic cat and other felines are definitive hosts, and a large number of mammals, including humans, and birds serve as potential intermediate hosts. Animals become infected when they ingest the infective stages of the parasite, which include mature (sporulated) oocysts containing sporozoites, pseudocytes containing bradyzoites, or free tachyzoites. The course of development of the parasite following its entry into the small intestine varies, and is dependent on whether a definitive or an intermediate host is infected. In cats and other felines parasite development involves an enteroepithelial cycle and an extraintestinal cycle. There is no enteroepithelial cycle in non-feline animals, which may serve as intermediate hosts (Fig. 2).

In the enteroepithelial cycle, some of the sporozoites liberated from oocysts in the lumen of the intestine of the cat penetrate the mucosal cells, become enclosed in a parasitophorous vacuole, and transform into trophozoites. The trophozoites reproduce asexually by schizogony producing numerous merozoites, which in turn infect other cells to repeat the asexual cycle. Eventually some merozoites undergo gamogony and give rise to sexual stages, (i.e., microgametocytes and macrogametocytes). Intracellular fertilization occurs when microgametes released from one cell enter other host cells containing macrogametes. Following fertilization, the resulting zygotes develop into immature oocysts that are voided with the feces. Peak oocyst production generally occurs approximately one week post-infection. Outside the host, sporulation occurs within one to five days, producing mature oocysts containing two sporocysts, each enclosing four infective sporozoites.

In the extraintestinal cycle, some of the sporozoites liberated in the small intestine enter the blood and are disseminated to various regions of the body, including the mesenteric lymph nodes, lungs, spleen, voluntary and cardiac muscles, retina, and cells of the nervous system. In these areas, the sporozoites invade cells and transform into rapidly dividing merozoites called tachyzoites. In acute infections there may be several cycles of cellular invasion, parasite multiplication, and host cell destruction. As the disease becomes chronic, infection is manifested by the formation of pseudocysts that envelope numerous parasites that are referred to as bradyzoites. Pseudocysts with infective bradyzoites may last for years in nerve tissue. Presumably pseudocyst formation is induced by host immune responses. As immunopathology decreases, the parasites are released from the cysts and infect other host cells. Thus, an alternation of proliferative and cystic phases occurs in infected intermediate hosts. Cats and other carnivorous animals readily acquire infections when they ingest oocysts or consume prey whose tissues contain pseudocysts or infective free parasites.

Human toxoplasmosis may result from either contact with cat feces contaminated with mature oocysts, or from ingesting infected meat or milk. Meat of various domestic animals may be infected, including pork, mutton, beef and poultry. Only asexual development of the parasite occurs in humans. Merozoites arising from asexual development enter the blood and lymphatics and form intracellular cysts in various tissues of the body. Symptoms vary greatly, with the majority of human infections benign or asymptomatic. The most serious are transplacental and neonatal infections. Clinical manifestations include hepatosplenomegaly, pneumonitis, retinochoroiditis, encephalomyelitis, cerebral calcification, hydrocephalus or microcephaly. Rarely is there complete recovery in infected children. The symptoms of postnatally acquired toxoplasmosis are frequently mild and may mimic those of infectious mononucleosis, with chills, fever, fatigue, headache, lymphadenitis, and myalgia (muscle pain).

It is estimated that approximately 13% of the world population is infected with *T. gondii.* In some European countries where raw meat commonly is eaten, the prevalence of human toxoplasmosis may be as high as 50%. In the United States approximately 4,000 infants are born annually with toxoplasmosis.

Genus *Plasmodium*

Species belonging to the genus *Plasmodium* are causative agents for malaria of humans and other animals in various tropical and subtropical regions of the world. The life cycle of these parasites involves an asexual phase (schizogony) alternating with a sexual one (gametogony), followed by sporogony. About forty species of female anopheline mosquitoes are definitive hosts, in which sexual reproduction occurs. Male mosquitoes do not feed on blood, and thus play no direct role in malarial transmission. Asexual reproduction occurs in the tissues of vertebrates, which are thus considered intermediate hosts. There are four species of malarial parasites infecting humans: *Plasmodium falciparum, P. vivax, P. malariae* and *P. ovale.*

Plasmodium falciparum is the cause of malignant tertian malaria also known as aestivo-autumnal or subtertian malaria. The parasite causes 50% of the malarial cases world wide. Falciparum malaria often terminates fatally. *Plasmodium vivax,* which causes benign tertian malaria, has the widest geographical distribution of the four species infecting humans. Indigenous cases of malaria occur as far north as England, Siberia, and Manchuria, and south into Argentina and South Africa. *Plasmodium malariae* is a relatively uncommon parasite producing quartan malaria. It is prevalent in tropical regions except South America. The parasite causes 7% of all malaria in the world. *Plasmodium ovale* is a rarely encountered species, but it has been reported from tropical and subtropical regions of many continents.

General Life Cycle of *Plasmodium*

A general account of the development of malarial parasites may be divided into three phases (Fig. 3):

1. Pre-erythrocytic or exoerythrocytic schizogony. This stage involves the development of merozoites from sporozoites introduced into humans when infected female mosquitoes bite. Within 30 minutes after infection, the sporozoites disappear from the peripheral circulation and invade parenchymal cells of the liver. Within the liver cells the parasites first develop into vegetative trophozoites. The parasites undergo one (*P. falciparum*) or two schizogonic cycles producing numerous exoerythrocytic merozoites. The merozoites escape from the liver cells and pass into the blood stream to invade erythrocytes. In *P. vivax, P. malariae* and perhaps *P. ovale,* exoerythrocytic trophozoites and merozoites may persist for years, causing relapse by producing numerous intermittent invasions of the bloodstream, especially in cases of declining immunity or ineffective drug therapy. Relapses do not occur in human infections with *P. falciparum* since exoerythrocytic multiplication is limited to a single generation of merozoites before invasion of the blood (Figs. 4A and 4B).

2. Erythrocytic schizogony. Merozoites that pass from the liver into the bloodstream invade erythrocytes and undergo schizogony. Within the red blood cells, growth of the parasite proceeds through the trophozoite and erythrocytic schizont stages. Infected erythrocytes enlarge as the intracellular parasites multiply. Eventually, the schizonts consist of numerous, fully formed merozoites. An erythrocytic schizont consisting of mature merozoites is termed a meront or segmenter. The segmenters

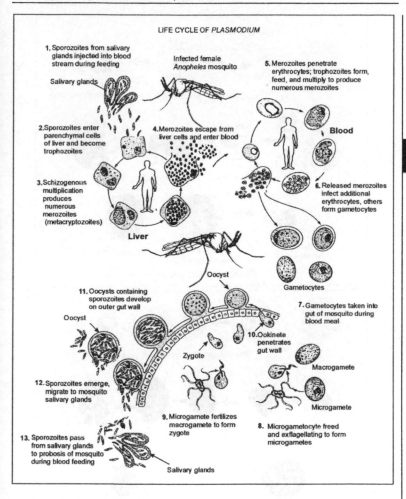

Figure 3. Life cycle of *Plasmodium*.

rupture almost synchronously, liberating the erythrocytic merozoites, which then invade other red blood cells repeating the asexual cycle. The cycle of schizogonic development and liberation of merozoites occurs typically every 48 hours in *P. vivax* and *P. ovale* infections, every 72 hours in *P. malariae* infections, and about every 36 to 48 hours in *P. falciparum* infections. The first visible manifestations of the disease occur with the synchronized lysis of red blood cells and the release of merozoites into the bloodstream, causing malarial paroxysms consisting of chills, burning fever, followed by sweating. Eventually, some erythrocytic merozoites develop into micro- and macrogametocytes instead of asexual schizonts.

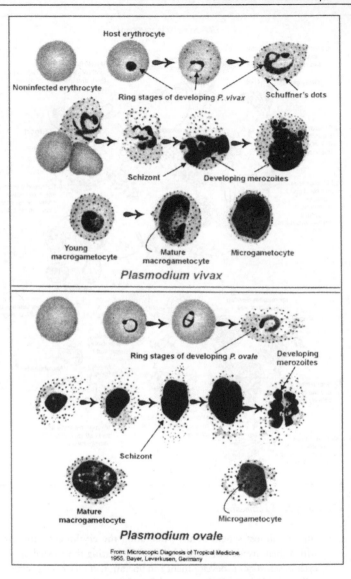

Plasmodium vivax

Plasmodium ovale

From: Microscopic Diagnosis of Tropical Medicine.
1955. Bayer, Leverkusen, Germany

Figure 4A. Intracellular developmental stages of *Plasmodium*. From Microscopic Diagnosis of Tropical Medicine. 1955. Bayer, Leverkusen, Germany.

Unfortunately, very little is known of the factors influencing gametogenesis. These immature gametocytes typically are present in circulating red blood cells after a few to several erythrocytic schizogonic cycles. They remain as gametocytes within the erythrocytes and do not mature further

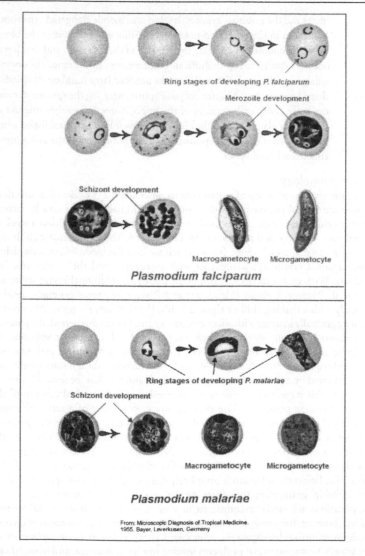

Ring stages of developing *P. falciparum*

Merozoite development

Schizont development

Macrogametocyte   Microgametocyte

Plasmodium falciparum

Ring stages of developing *P. malariae*

Schizont development

Macrogametocyte   Microgametocyte

Plasmodium malariae

From: Microscopic Diagnosis of Tropical Medicine.
1955. Bayer, Leverkusen, Germany.

Figure 4B. Intracellular developmental stages of *Plasmodium*. From Microscopic Diagnosis of Tropical Medicine. 1955. Bayer, Leverkusen, Germany.

in the vertebrate intermediate host. The remaining sexual phase of the life cycle, involving fertilization and sporogony, occurs in the female mosquito.

3. Sexual phase. Further development of the gametocytes occurs in the gut of the female mosquito where the parasites are liberated from digested erythrocytes taken in with a blood meal. Microgametes fertilize macroga-

metes and the resulting zygotes develop into mobile elongated organisms, 15-20 μm in length, called ookinetes. Within a few days after the blood meal, the ookinetes penetrate the gut wall of the mosquito and develop as oocysts between the epithelium and the basement membrane. The oocysts enlarge rapidly, and within 2-3 weeks produce large numbers of spindle-shaped sporozoites. Mature oocysts rupture, releasing the sporozoites into the body cavity or hemocoel of the insect. Some sporozoites migrate to the salivary glands, which they invade. A new infection is established when the insect next feeds and sporozoites are introduced with the saliva into a cutaneous blood vessel.

Symptomology

A primary clinical attack of malaria has its onset about 2-3 weeks after infection. Seldom are typical paroxysms evident during the initial stages of attack, instead patients exhibit sustained or irregularly remittent fever. However, within a week of the primary attack typical paroxysms are experienced. The characteristic chills and fever of a malarial attack result from the release into the blood of necrotic debris from ruptured erythrocytes, together with merozoites and their metabolic by-products. With each successive schizogony, numerous additional erythrocytes are destroyed. Primary *P. vivax, P. malariae* and *P. ovale* infections generally develop suddenly with a shaking chill or rigor. Initially, the chill lasts for up to 20 minutes, and may gradually increase with successive paroxysms. During this period, the patient's temperature is actually elevated. High fever follows and may be accompanied by headache, muscular and abdominal pain, anorexia, nausea, vomiting, and increased pulse and respiratory rates. After continuing for several hours, the hot stage terminates, and is followed by a profuse sweating stage, which also may last for several hours. At the end of this stage, the patient is usually weak, but feels marked relief until the onset of the next paroxysm. The primary attack, consisting of a series of several paroxysms, may extend over a period of a month or more before the parasites completely disappear from the blood and the symptoms are terminated. As the primary attack wanes, paroxysms frequently become less severe and irregular in periodicity before they disappear. Relapses of vivax malaria may continue for a period of five years before the infection is completely eliminated. Relapses of quartan malaria may develop years after the onset of the initial malarial paroxysm. Serious complications are rarely encountered in vivax and quartan malaria. Falciparum malaria, however, frequently produces severe complications. Manifestations of cerebral involvement include hyperpyrexia, convulsions, coma, and death resulting from shock and anoxia. Gastrointestinal problems involve vomiting, diarrhea, and hemorrhage. Lysis of erythrocytes plus enhanced phagocytic activity results in anemia and enlargement of the spleen and liver. The liver may be congested and contain deposits of pigment (hemozoin) derived from hemoglobin of infected erythrocytes. In severe infections, the number of erythrocytes may be reduced by 20%. Infected erythrocytes may form multiple thrombi in various capillaries. The capillaries of the lungs, brain, and kidneys frequently become thrombotic with accumulations of infected erythrocytes, pigment and macrophages, and may rupture producing multiple extravasations.

Piroplasmea

The class Piroplasmea is comprised of a single order, Piroplasmida, whose members are of considerable veterinary importance. Apparently, the organisms reproduce only asexually, by binary fission or schizogony, in the blood cells of vertebrates. Infection is transmitted by various species of ticks, in which a sexual multiplication cycle occurs.

Genus Babesia

Human babesiosis is a fulminating hemolytic disease similar to malaria. Of the approximately 25 cases reported in humans, two have been attributed to *Babesia divergens*, a parasite of cattle, and the others to *B. microti*, a parasite of wild rodents, which can also infect dogs and cats. Most of the *B. microti* infections in humans have been reported from Nantucket Island, Martha's Vineyard, Shelter Island (near Long Island, New York), and Montauk, Long Island, New York. In the mammalian host, the parasites occur only in the trophozoite stage in the erythrocytes where they multiply by binary fission. Upon destruction of the blood cells, the parasites escape and invade other host erythrocytes continuing the cycle. Ticks become infected when they feed on the blood of an infected vertebrate and ingest the intraerythrocytic parasites, but they do not themselves transmit the disease. Instead, the parasites penetrate the gut of the tick and enter the eggs developing in the ovaries, thus infecting the young ticks, which eventually hatch from these eggs. Within the young ticks, the parasites migrate to the salivary glands, and are injected into the vertebrate host by the feeding tick. The transmission of parasites from an infected female to her offspring through the eggs is referred to as transovarian transmission (Fig. 5).

Clinical manifestations of babesiosis are noted within 10 days after infection and may last for several weeks. The illness, which mimics malaria, is characterized by fever, chills, drenching sweats, myalgias, fatigue, weakness, hemolytic anemia, and enlargement of the spleen and liver. Death results from the accumulation of toxic metabolites and anoxia from capillary occlusions.

Other Apicomplexa

Genus Pneumocystis

Pneumocystis carinii causes is an extracellular parasite found in the interstitial tissue of the lung where it causes interstitial plasma cell pneumonia. Only an intrapulmonary cycle is known for this parasite. The parasite has a cosmopolitan distribution and is found in infants, in patients with certain immunologic disorders such as leukemia, Hodgkin's disease, and hypo- or agammaglobulinemia, and in individuals undergoing immunosuppressive therapy. The organism occurs as a uninucleate pleomorphic trophozoite with a doubly contoured outer membrane, or as a cyst-like form containing as many as eight intracystic sporozoites. The intracystic sporozoites rupture from the cyst and develop into haploid trophozoites. The haploid trophozoites combine to form diploid trophozoites. Asexual development follows giving rise to precyst and mature cyst forms.

In young children, the mortality rate attributed to *P. carinii* infections may range from 30 to 40 percent. The organism is believed to be present in many humans, but may be incapable of producing a disease in healthy, immune competent hosts. The incubation period ranges from two to eight weeks. The disease is considered highly contagious. It is believed that the mode of transmission among humans is by

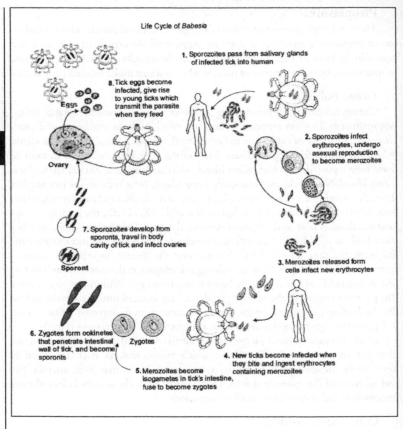

Life Cycle of *Babesia*

1. Sporozoites pass from salivary glands of infected tick into human

8. Tick eggs become infected, give rise to young ticks which transmit the parasite when they feed

2. Sporozoites infect erythrocytes, undergo asexual reproduction to become merozoites

7. Sporozoites develop from sporonts, travel in body cavity of tick and infect ovaries

3. Merozoites released form cells infect new erythrocytes

6. Zygotes form ookinetes that penetrate intestinal wall of tick, and become sporonts

4. New ticks become infected when they bite and ingest erythrocytes containing merozoites

5. Merozoites become isogametes in tick's intestine, fuse to become zygotes

Figure 5. Life cycle of *Babesia*.

inhalation of cysts, and by transplacental transmission. The clinical course of the disease is marked by progressive dyspnea, cyanosis, tachypnea, and infiltration of the lungs, with death resulting from asphyxia. The recent increase in the incidence of pneumonocystis pneumonia is coincident with the occurrence of AIDS. Pneumocystis is the most common cause of death among AIDS patients, affecting about 60% of the patients with this immune deficiency disease.

Cryptosporidium parvum

The coccidian *Cryptosporidium parvum* causes cryptosporidiosis, a common gastrointestinal disorder. The disease, which occurs in a wide variety of hosts including primates, cattle, rodents and birds, is of clinical importance in immune compromised individuals. Symptoms include diarrhea, nausea, abdominal pain, vomiting, biliary disorder, and pneumonitis. In immune compromised individuals, infections may last several months and terminate in death. Human infection results from the ingestion of oocysts present in contaminated food or water. Sporozoites excyst in the intestine and invade the epithelial cells of the gut and respiratory system. Sporozoites develop into feeding trophozoites seen within parasitophorous vacuoles. The merozoites that are produced by asexual reproduction escape and invade nearby cells to repeat the infection cycle. Ultimately, microgametocytes and macrogametocytes are produced. These differentiate into gametes that fuse to form oocytes that are voided with the host feces.

4

CHAPTER 5

Digenetic Trematodes: Flukes

Digenetic trematodes comprise a large group of endoparasites variable in size, shape, and habitat. The body of most adult flukes is flattened dorsoventrally, and covered with a smooth or spiny, resistant cuticle. Adult flukes inhabit the digestive tract, bile passages, lungs, or blood of their vertebrate hosts.

Adult digenetic flukes have two suckers or attachment organs, an anterior oral sucker surrounding the mouth, and a more posterior ventral sucker or acetabulum. The mouth or buccal cavity leads to a muscular pharynx and esophagus, and then into branched cecae (Fig. 1). The cecae run parallel to each other and end blindly, either in the posterior portion of the worm, or about halfway down the body. The parasites feed on mucus, blood, and other host tissues. Except for schistosomes or blood flukes, the trematodes infecting humans are hermaphroditic, and also capable of self-fertilization. Typically, the male reproductive system consists of two testes, seminal ducts (vas efferens) that connect to form a vas deferens, seminal vesicle, ejaculatory duct, and a muscular cirrus or copulatory organ that terminates at the male genital pore located within a common genital atrium. The genital atrium, which is commonly found on the midventral surface and anterior to the acetabulum, opens to the exterior via a gonopore. The cirrus may be enclosed within a cirrus pouch, and evaginated for sperm transfer to the female system. The female reproductive system consists of a single ovary, and oviduct, seminal receptacle, an ootype or egg chamber, and a uterus that extends to the female genital pore. The female genital pore is usually located near the male genital pore within the genital atrium. Self- and cross-fertilization are possible modes of reproduction. Following copulation, spermatozoa are stored in the seminal receptacle and released as required.

Life Cycle

The typical life cycle of a digenetic trematode is comprised of an asexual phase in molluscs, usually gastropods, and a sexual phase in vertebrates, the definitive hosts. Many species have a second intermediate host in which the sexual phase begins. The asexual phase includes an egg that is either unembryonated when discharged from the definitive host, or it is embryonated and contains a fully formed ciliated larval stage called a miracidium (Fig. 1). Most eggs have an operculum or cap at one end through which the miracidium emerges. In the majority of species, further development is possible when the miracidium hatches and penetrates the soft tissues of a suitable species of snail. In other cases, the eggs do not hatch outside the definitive host, but rather are ingested by the intermediate host. It is possible that miracidia respond to various conditions, including light, temperature, salinity, pH, and gravity, that bring the parasites into the range of a suitable molluscan host. There is some evidence that some miracidia are attracted to their molluscan host by chemotactic stimuli. Penetration of the host is affected by a boring action of the larva, a process

Parasites of Medical Importance, by Anthony J. Nappi and Emily Vass.
©2002 Landes Bioscience.

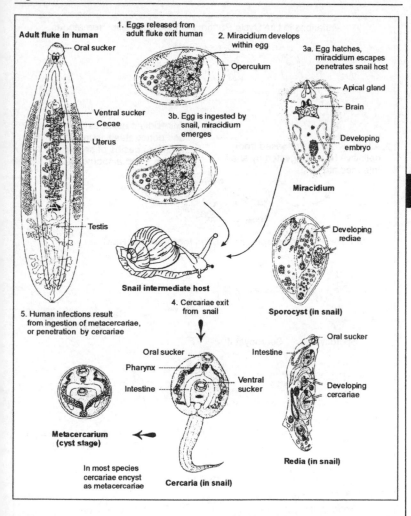

Figure 1. Developmental stages of digenetic trematodes. Modified from Hegner RW, Engemann. Invertebrate Zoology. New York: The Macmillan Company, 1968.

that is facilitated by the release of proteolytic enzymes from glands located in the apical region of the body of the miracidium.

Within the miracidium, a mass of germinal tissue provides the cell lineage for succeeding stages in the life cycle. Upon entering the molluscan intermediate host, the miracidium sheds its ciliated covering, elongates and transforms into either a saclike first-generation (mother) sporocyst, or in some species a slightly more advanced stage called a redia. Germ cells within the body of the mother sporocyst or redia develop into either second generation (daughter) sporocysts or rediae. Germinal cells in the daughter sporocysts or rediae eventually give rise to cercariae, the final

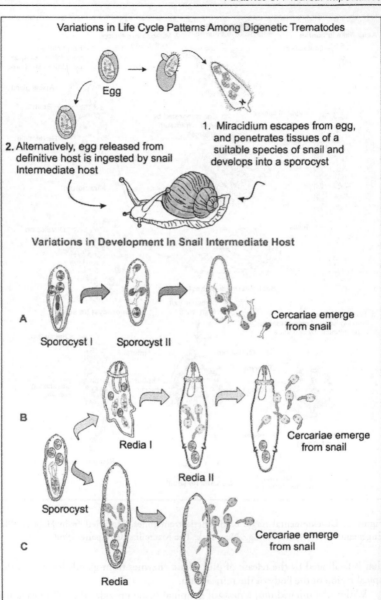

Figure 2. Variations in the life cycle of digenetic trematodes.

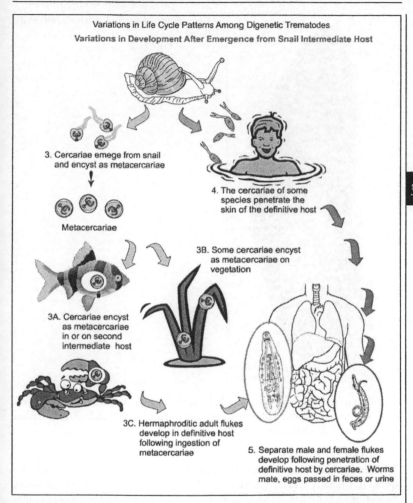

Variations in Life Cycle Patterns Among Digenetic Trematodes
Variations in Development After Emergence from Snail Intermediate Host

3. Cercariae emege from snail and encyst as metacercariae

Metacercariae

4. The cercariae of some species penetrate the skin of the definitive host

3A. Cercariae encyst as metacercariae in or on second intermediate host

3B. Some cercariae encyst as metacercariae on vegetation

3C. Hermaphroditic adult flukes develop in definitive host following ingestion of metacercariae

5. Separate male and female flukes develop following penetration of definitive host by cercariae. Worms mate, eggs passed in feces or urine

Figure 3. Variations in the life cycle of digenetic trematodes.

product of the asexual phase in the molluscan host. The cercariae are essentially young flukes. They possess a mouth, gut, suckers, penetration glands (secretions from which facilitate the invasion of the molluscan intermediate host), and/or cystogenous glands (whose secretions may provide a temporary protective layer). During their development, propagatory cells, derived from the germ cells, produce the stem cells of the adult reproductive system. In most species, the cercariae escape from the first intermediate host and encyst on objects in the water, or in the body of a second intermediate host, and then transform into metacercariae which develop and become infective to the definitive host. The definitive host becomes infected when it ingests the plant or animal harboring the metacercariae. Species specific

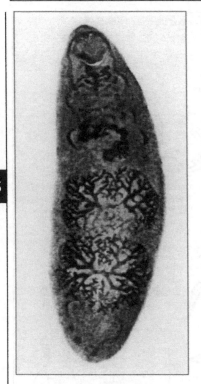

Figure 4. *Fasciolopsis buski* adult, an intestinal fluke.

5

variations in the life cycle pattern include: (1) more than one generation of sporo-cysts or rediae; (2) deletion of either sporocyst or redial stages; and, (3) penetration by cercariae of the definitive host (Figs. 2 and 3).

Intestinal Flukes

Fasciolopsis buski is an intestinal parasite of pigs and humans. This fluke is wide-spread in China, Vietnam, Thailand, and Indonesia. The parasite is a large worm, measuring up to 8 cm in length (Fig. 4). Each worm may produce as many as 25,000 eggs per day.

Infection with *F. buski* is acquired by ingestion of metacercariae encysted on freshwater edible plants such as bamboo shoots, water chestnut, water hyacinth, and water caltrop (Fig. 5). After excystation in the duodenum, the larvae attach to the intestinal mucosa and in about three months develop into adult parasites. Embryonation occurs in fresh-water after the eggs pass with the feces from the host. Hatched miracidia penetrate the soft tissues of certain planorbid snails (*Segmentina hemispaerula*, *S. trochoides*, *Hippeuitis contori*, *Gyraulus* sp.) and develop into sporo-cysts. Two redial generations are followed by the formation of numerous cercariae, which pass from the snail and after swimming about, move onto aquatic vegetation and encyst. At the site of attachment, these large flukes produce a local inflamma-tion and ulceration, occasionally accompanied by hemorrhage. In severe infections,

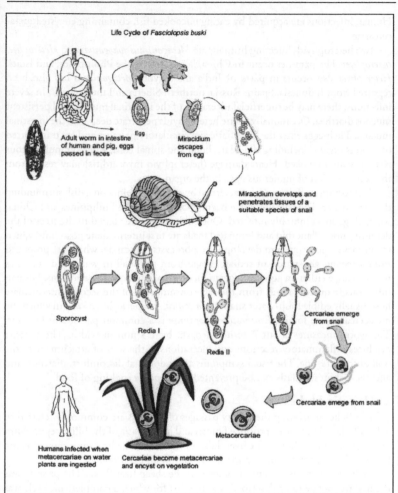

Life Cycle of *Fasciolopsis buski*

Adult worm in intestine of human and pig, eggs passed in feces

Egg

Miracidium escapes from egg

Miracidium develops and penetrates tissues of a suitable species of snail

Sporocyst

Redia I

Redia II

Cercariae emerge from snail

Cercariae emege from snail

Metacercariae

Humans infected when metacercariae on water plants are ingested

Cercariae become metacercariae and encyst on vegetation

Figure 5. Life cycle of *Fasciolopsis buski*.

there may be acute intestinal obstruction, abdominal pain, diarrhea, anorexia, nausea, vomiting, edema of the face, abdomen, and lower extremities, and yellow profuse stools indicative of malabsorption.

Heterophyid flukes are small, ovoid-shaped flatworms that live attached to the intestinal mucosa of fish-eating birds and mammals. The flukes are generally less than 3 mm in size, and produce operculate eggs. Hatching does not occur unless the eggs are ingested by appropriate species of freshwater snails. Within the snail, a single sporocyst generation is followed by two generations of rediae, and then cercariae. After escaping from the snail, the cercariae encyst as metacercariae either on the underside of the scales of certain fresh-water fishes or in the muscles of the fish.

Human infections are acquired by eating uncooked fish containing encysted metacercariae.

Two heterophyids infecting humans are *Metagonimus yokogawai* and *Heterophyes heterophyes*. The parasites occur in China, Japan, Korea, the Philippines, and Israel. *Heterophyes* also occurs in parts of India and Egypt. *Metagonimus* has also been reported from Indonesia, Spain, Russia, northern Siberia and the Balkans. In severe infections, there may be superficial ulceration of the intestinal mucosa and persistent mucous diarrhea. Occasionally, some heterophyids penetrate deep into the intestinal mucosa. Their eggs enter the lymphatic and circulatory passages and are transported to various organs, including the heart, brain and spinal cord, where granulomatous responses are provoked. Heart damage (heterophyid myocarditis) may result from the accumulation of numerous eggs in the organ.

Echinostomate flukes are characterized by a collar of spines on a disk surrounding the oral sucker. The parasites have been reported from the Philippines and China, where dogs are commonly infected. Operculated eggs are voided in the feces of the definitive host. Planorbid and lymnaeid snails are first intermediate hosts, into which the miracidia penetrate and develop into sporocysts. Cercariae, which are produced from a second generation of rediae, escape from the snail in water and encyst as metacercariae either on aquatic vegetation, or in various aquatic organisms. Human infections frequently result from the raw consumption of the second intermediate host, usually edible fresh-water snails (*Pila conica, Viviparus javanicus*), in which the metacercariae are found. *Echinostoma ilocanum* is a common parasite of humans. The worm measures about 7 mm in length, by 1.5 mm in width. The parasite produces inflammatory reactions and ulcerations at the sites of attachment to the wall of the intestine. The usual symptoms are abdominal discomfort, diarrhea, and anemia. Echinostomiasis can be prevented by adequate cooking of food.

Hepatic Flukes

The flukes inhabiting the biliary passages of humans are commonly referred to as liver flukes. The worms produce fibrosis and hyperplasia of the biliary epithelium leading to portal cirrhosis. The liver flukes of significant medical importance belong to the following three genera: *Fasciola, Clonorchis* and *Opisthorchis*.

Fasciola hepatica is commonly known as the sheep liver fluke. The parasite was the first trematode to be described, and the first for which a complete life cycle was elucidated. The parasite has a widespread distribution throughout many sheep- and cattle-raising countries of the world. Numerous human infections have been reported from Latin America, France, Algeria, England, Germany, Poland, China, Russia, Hawaii, parts of Africa, and in some areas of the southern continental United States.

The adult fluke is flattened and leaf-shaped, with an anterior conical projection and a broadly rounded posterior. The adults, which measure up to 3 cm in length and 1.3 cm in width, produce operculate eggs. Human fascioliasis is acquired by ingesting vegetation, usually freshwater cress, on which encysted metacercariae are found. The metacercariae excyst in the gut, and the parasite penetrates the intestinal wall, migrates through the peritoneal cavity and enters the bile ducts and liver parenchyma. Occasionally, some adult flukes wander in the peritoneal cavity and other ectopic areas producing necrotic foci with fibrosis. While grazing, infected sheep and cattle contaminate vegetation and water sources with feces containing

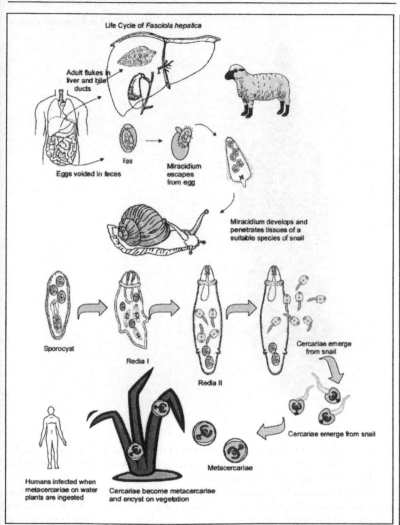

Figure 6. Life cycle of *Fasciola hepatica*.

fluke eggs. Upon hatching, the miracidia have about 24 hours in which to find a suitable lymnaeid snail host. A miracidium penetrates the snail, sheds its ciliated epithelium, and transforms into a sporocyst. Two generations of rediae are followed by cercariae, which begin emerging from the snail 5-7 weeks post infection. The cercariae attach to any available object and transform into metacercariae (Fig. 6).

Severe infections may be characterized in part by mechanical obstruction, irritation, and inflammation of host tissues. Frequently, there are hyperplastic changes in

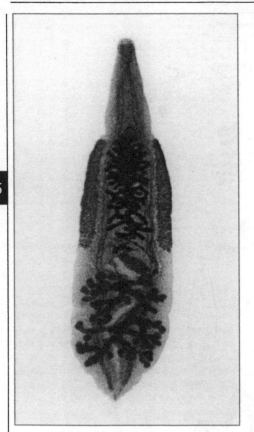

Figure 7. Adult *Clonorchis sinensis*, a hepatic fluke.

5

the biliary epithelium, fibrosis and/or erosion of the walls of the bile ducts, atrophy of the liver parenchyma, and cirrhosis. Migrating larvae may produce lesions in the eye, brain, lungs, skin and other ectopic foci.

Clonorchis sinensis, the Chinese liver fluke, occurs in China, Taiwan, Japan, Korea, and Vietnam. Dogs and cats are important reservoir hosts. Adult worms are elongate and flat, measuring as much as 2.5 cm in length by 0.5 cm in width (Fig. 7). They occupy the bile ducts, most frequently in the more distal regions, just under the surface of the liver. The eggs are discharged into the biliary passages and pass in the feces of the host. The average daily production of eggs per worm is about 2000. The eggs hatch when ingested by a suitable species of operculate snails (*Bulimus, Alocinma* and *Parafossarulus*), and the miracidia transform into sporocysts. A single generation of rediae is followed by the production of cercariae. After emerging from the snails the cercariae penetrate the tissues of certain fresh-water fishes and become encysted. Human infection is acquired when uncooked fish containing the encysted metacercariae is eaten (Fig. 8). Upon excystation in the duodenum, the larvae enter the smaller biliary passages by migrating through the ampulla of Vater.

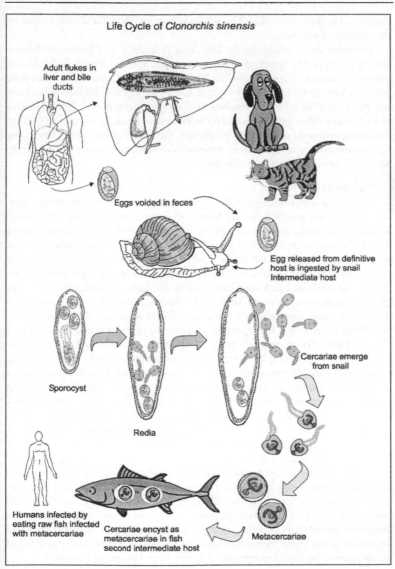

Life Cycle of *Clonorchis sinensis*

Adult flukes in liver and bile ducts

Eggs voided in feces

Egg released from definitive host is ingested by snail Intermediate host

Cercariae emerge from snail

Sporocyst

Redia

Humans infected by eating raw fish infected with metacercariae

Cercariae encyst as metacercariae in fish second intermediate host

Metacercariae

Figure 8. Life cycle of *Clonorchis sinensis*.

Chronic clonorchiasis is generally asymptomatic. In severe infections, the acute period lasts less than a month. The infection may be characterized by chills, fever, epigastric pain, diarrhea, enlargement and tenderness of the liver, congestive splenomegaly, hyperplasia of bile duct mucosa, biliary obstruction, cholangitis, and liver abscesses. The degree to which these symptoms are manifested is directly related to the intensity of infection.

Two other flukes inhabiting the bile ducts of humans are *Opisthorchis felineus* and *O. viverrini*. The symptoms of opisthorchiasis are very similar to the disease picture produced by *Clonorchis sinensis*. Except for different species of intermediate hosts, their life cycles are similar to that of *C. sinensis*. *Opisthorchis felineus* has been reported from Siberia, Poland, southeastern Europe, Asia, and India. Several species of fresh-water fishes serve as secondary intermediate hosts. Dogs, cats, and other fish-eating mammals are naturally parasitized. The snail *Bulimus tentaculatus* is the first intermediate host. *Opisthorchis viverrini* is endemic to northeastern Thailand, where fish-eating mammals are affected.

Pulmonary Flukes

Flukes inhabiting the lungs of various mammals belong to the genus *Paragonimus*. The best known species infecting man is *P. westermani*, the cause of endemic hemoptysis or paragonimiasis in regions of Thailand, Indonesia, China, Central and South America, Africa, and the South Pacific. The adult worms, which measure up to 1.6 cm in length and 0.8 cm in width, are characteristically found enclosed in cystic structures near the bronchi. The eggs are coughed up and expectorated in the sputum, or swallowed and later evacuated in the feces. In fresh-water, the eggs embryonate in about two weeks, then hatch. The free swimming miracidia penetrate appropriate snails (*Semisulcospira libertina, Melania amurensis, M. obliquegranosa* and *Brotia asperata*) and transform into sporocysts, in which rediae are developed (Fig. 9). Each redia subsequently produces a brood of cercariae. Emerging cercariae next invade the tissues of suitable fresh-water crustaceans, especially crabs or crayfish (*Cambarus, Pseudotelphusa, Potamon, Paratelphusa* and *Eriocheir*), and form metacercariae. Human infection results from eating raw crabs or crayfish containing encysted metacercariae. Following excystation in the duodenum, young worms penetrate the intestinal wall, migrate through the peritoneal cavity, pass through the diaphragm and enter the peribranchial tissues. The circuitous migratory route through the body cavity may take about three weeks before the worms enter the lungs, in which they develop into adults in 5-6 weeks.

Paragonimus westermani has been found in many abnormal or ectopic sites, such as various muscles, heart, liver, brain, mesenteric lymph nodes, testes, and pleural or peritoneal cavity. In these locations, fibrous inflammatory cysts are formed around the worms. Some of these lesions may be ulcerative. Cerebral paragonimiasis frequently results in intracranial calcification, impaired vision, and epilepsy. Lung infections may be characterized by occasional coughing with the discharge of blood in the sputum (periodic hemoptysis), dyspnea, fever, fatigue, and anorexia.

Blood Flukes

Trematodes belonging to the genus *Schistosoma* inhabit the circulatory system of humans causing one of the more serious diseases of helminth origin, schistosomiasis (bilharziasis). The schistosomes differ from most other trematodes in that the body is cylindrical, and the sexes are separate. The worms exhibit sexual dimorphism, with females slightly longer and more slender than males. Males possess a ventral sex canal (gynecophoric canal) in which the female reposes (Fig. 10). Adult schistosomes

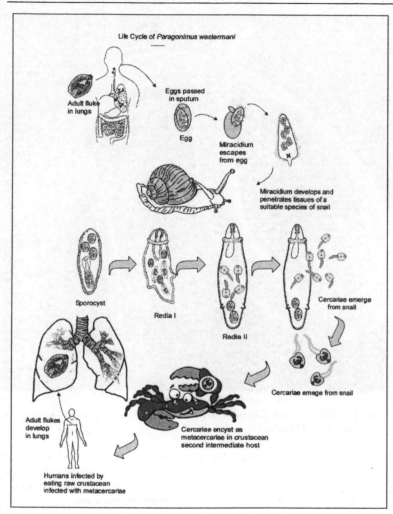

Figure 9. Life cycle of *Paragonimus westermani*.

characteristically live in pairs in the portal venous blood vessels or in the vesicle venules of the caval system, where females lay non-operculate, partially embryonated eggs. The worms live on the average of a few years, but in exceptional cases they can persist for as long as 30 years or more. The females produce prodigious numbers of eggs, which are forced through the walls of either the intestine or urinary bladder and are discharged in the excreta. The eggs hatch in fresh-water and the miracidia penetrate suitable species of snails, in which they develop into sporocysts (Fig. 11). The cercariae, which possess a forked tail, are produced from second generation sporocysts. The cercariae escape from the snail and, on contact with

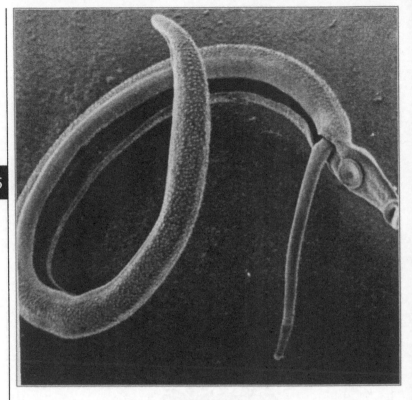

Figure 10. Male and female *Schistosoma mansoni* in copula. Scanning electron micrograph courtesy of Harvey Blankespoor.

human skin, penetrate the body (Fig. 12), enter the cutaneous blood vessels and initiate infection. Young flukes are first transported to the liver sinusoids where they feed and grow for a period of 5 to 6 weeks. When mature, they migrate into the portal system to their final location in mesenteric or vesicular veins.

Three species of schistosomes infect humans:

1. *Schistosoma mansoni* occurs over extensive areas of Africa, Egypt, the Arabian peninsula, and parts of South America and the West Indies. Various primates, insectivores, rodents, and marsupials are also infected with this fluke. Male parasites measure up to 1 cm in length, females about 1.6 cm. The eggs are elongate and oval and possess a conspicuous lateral spine projecting near one pole (Fig. 13). Adult worms are usually found in smaller branches of the inferior mesenteric vein, intrahepatic portal blood, vesicle venules, and pulmonary arterioles. The pulmonate snail *Biomphalaria glabrata* is the intermediate host of *S. mansoni*.

2. *Schistosoma haematobium* occurs in Egypt, Africa, Malagasy, Arabian peninsula, Iran, India, Israel, Portugal, Cyprus, Syria, and Lebanon. Male

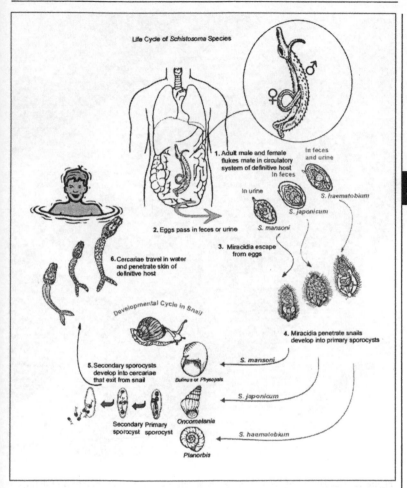

Figure 11. Life cycle of *Schistosoma*.

worms may attain a length of 1.5 cm, and females a length of 2 cm. The eggs have a conspicuous terminal spine (Fig. 13). The adult worms occupy the blood vessels surrounding the bladder. The parasite is carried by pulmonate snails of the genus *Bulinus*.

3. *Schistosoma japonicum* occurs in Japan, Korea, Formosa, China, Taiwan, Laos, Cambodia, Thailand, the Philippines, and Celebes. Male worms may attain a length of 2.2 cm and females 3.0 cm. The spherical eggs are smaller than those of the other two schistosome species and possess a minute, blunt projection may be seen on some eggs near one pole (Fig. 13). Adult worms live within the branches of the superior mesenteric

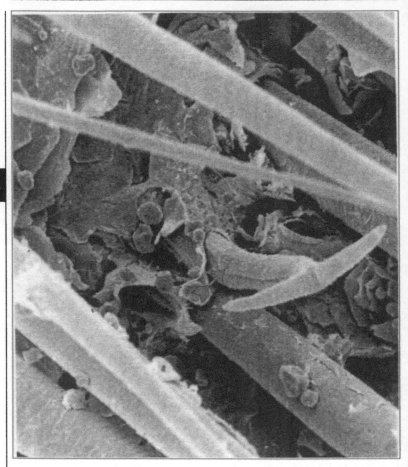

Figure 12. Caudal end of *Schistosoma mansoni* cercaria penetrating epidermis of definitive host shortly after emergence from snail intermediate host. Scanning electron micrograph courtesy of Harvey Blankespoor.

vein, near the small intestine. As the disease progresses, the inferior mesenterics and the caval system may also be invaded. Gill-breathing snails of the genus *Oncomelania* are the principal intermediate hosts for *S. japonicum*.

Early symptoms of schistosomiasis include localized dermal hemorrhages, edema and pruritus at the site of penetration. These symptoms typically disappear in less than one week. During the succeeding month, fever, toxic and allergic manifestations, accompanied by abdominal distress may develop. Migration of the parasites through the lungs may cause spasmodic cough, hemoptysis, and severe pleuritic chest pain. Progressive pathologic conditions include bronchitis, fibrosis, and pleu-

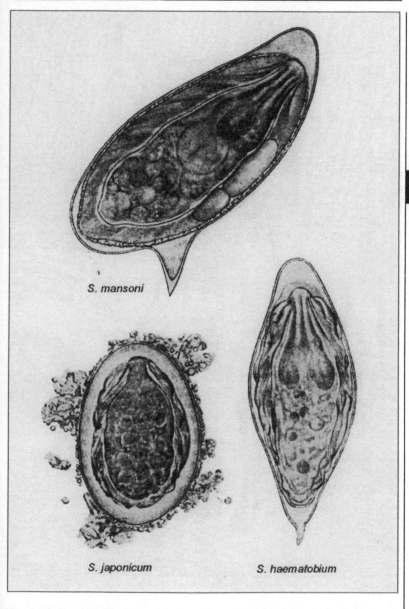

S. mansoni

S. japonicum

S. haematobium

Figure 13. Eggs of *Schistosoma* species. From Microscopic Diagnosis of Tropical Medicine. 1998. Bayer, Leverkusen, Germany.

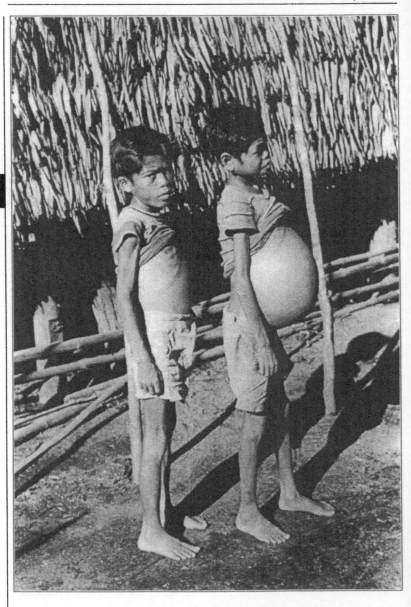

Figure 14. Ascites in advanced schistosomiasis japonica (Leyte, Philippines). This is an example of dwarfing (individual at right) caused by the parasite. The male of the left is 13 years old, the individual on the right is 24 years old. Photograph courtesy of Robert E. Kuntz.

ral effusion. Acute hepatitis is generally manifested soon after the parasites invade the liver. Other symptoms appearing during advanced stages of infection include fever, uticaria, epigastric discomfort, diarrhea, dysentery, the accumulation of fluid in the abdominal cavity (ascites) (Fig. 14), hepatosplenomegaly and liver dysfunction. Hematuria (blood urine), cystitis (inflammation of the bladder), and dysuria are common symptoms in patients infected with *S. haematobium*. There may be extensive tissue damage due to the extrusion of eggs through the wall of the intestine or urinary bladder. Eggs lodged in various tissues provoke inflammatory reaction, with leukocytic and fibroblastic infiltration. Fibrous nodules (granulomas) enclosing small accumulations of eggs are frequently formed on the serosal and peritoneal surfaces. Necrosis, ulceration, and malignant changes may follow. Neurological pathologies, involving coma and paralysis, result from parasite eggs invading the brain.

Schistosome dermatitis or "swimmer's itch" in humans results from the penetration of the skin with non-human (i.e., bird, cattle, rodent) schistosome cercariae in fresh and brackish water. Fortunately, the parasite does not produce a permanent infection. Infection produces an initial severe prickling, local edema, generalized urticaria, and the development of macules and pustules. Schistosome dermatitis has been reported from many regions of the continental United States, Hawaii, Europe, Latin America, India, and Thailand.

CHAPTER 6

Cestodes

Cestodes or tapeworms are parasitic during all or a major portion of their lives. With few exceptions, adult cestodes possess an elongated tape-like body, and they lack a digestive tract. The latter feature separates these worms from trematodes and nematodes. Typically, the habitat of the adult tapeworm is the intestinal tract of its host. Cestode larvae, however, invade a wide range of host tissues, although most larvae demonstrate a preference for particular, species-specific, sites.

The body of a tapeworm consists of an anterior attachment organ or scolex, followed by an unsegmented neck, succeeded by a chain of proglottids ("segments") termed the strobila (Fig. 1). The number of proglottids may vary from 3 to 4 in the hydatid worm (*Echinococcus granulosis*), to more than four thousand in the broad or fish tapeworm (*Diphyllobothrium latum*). The scolex may be equipped with various holdfast organs, which secure the worm to the mucosa of the host's small intestine. There are essentially three types of adhesive structures: (1) Bothridia are broad, leaf-like structures with flexible margins. They usually occur in groups of four and project from the dorsal or ventral side of the scolex. (2) Bothria are dorsal or ventral grooves of weak muscularity. Usually two are found on the scolex. (3) Acetabula are highly muscular, cup-shaped adhesive structures (suckers). There are frequently four acetabula on a scolex. With the exception of the broad tapeworm, which possesses bothria, all tapeworms of humans have four cup-shaped suckers on the scolex. In addition to suckers, most tapeworms have keratinaceous hooks that anchor the scolex to the intestinal wall. In acetabulate cestodes, the hooks may be arranged circularly on a protrusible cone-like structure, the rostellum. In some cases, the rostellum lacks hooks, and is termed "unarmed".

In many cestodes, the strobila grows continuously throughout the life of the worm by asexual budding (strobilization) of new proglottids in the neck region. Each proglottid moves posteriorly as a new one is formed. As new proglottids are added, the strobila elongates so that in some species enormous lengths are attained. As the proglottids move from the neck region, the reproductive organs mature and the eggs are fertilized. The most recently formed and immature proglottids are found nearest the scolex, while the larger, mature proglottids are found near the middle of the strobila. The terminal portion of the strobila contains ripe or gravid proglottids filled with eggs. Typically, each proglottid contains one or more sets of reproductive organs. Mature segments contain both male and female reproductive organs, and thus are capable of self fertilization. Cross fertilization also is possible between different segments of a single worm, or between segments of two worms living together within the host. Usually the male organs mature first and produce sperm that are stored until eggs are manufactured (protandry or androgyny). In some species, the ovary matures before the testes (protogyny or gynandry).

Parasites of Medical Importance, by Anthony J. Nappi and Emily Vass.
©2002 Landes Bioscience.

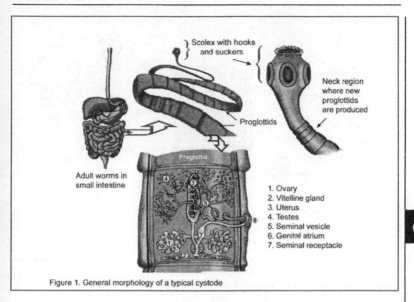

Scolex with hooks and suckers

Neck region where new proglottids are produced

Proglottids

Proglottid

Adult worms in small intestine

1. Ovary
2. Vitelline gland
3. Uterus
4. Testes
5. Seminal vesicle
6. Genital atrium
7. Seminal receptacle

Figure 1. General morphology of a typical cystode

Figure 1. Morphology of a typical cestode.

Developmental Stages and Life Cycles

There are two basic life cycle patterns among the cestodes parasitizing humans; one exhibited by pseudophyllideans, the other by cyclophyllideans. In both, the egg gives rise to an oncosphere or hexacanth larva, so-called because it possesses six elongated hooks at the posterior pole. An egg membrane or embryophore surrounding the embryo may be ciliated, in which case the parasite is termed a coracidium. Virtually every known tapeworm has an indirect life cycle involving at least one intermediate host that ingests the eggs that pass from the intestinal tract of the definitive host. One notable exception is *Vampirolepis (Hymenolepis) nana*, a cyclophyllidean with a direct life cycle in which both larval and adult stages occur in the definitive host (either mice or humans).

Pseudophyllidean Life Cycle

The eggs of pseudophyllids are generally unembryonated when shed from the definitive host, but in water they soon embryonate and hatch. The oncosphere emerging from the egg has a ciliated embryophore and is called a coracidium (Fig. 2). Within the first intermediate host the coracidium sheds its ciliated epithelium and actively penetrates the gut. Within the body cavity it develops into an elongated, larva called a procercoid. The embryonic hooks are retained and, in some species, are borne on a posterior structure termed a cercomer. However, the definitive holdfast has not yet developed. When the first intermediate host is consumed by the second intermediate host, usually a freshwater fish, the larva penetrates the gut and migrates into the connective tissue and skeletal muscles, and transforms into the last larval stage, the plerocercoid. The embryonic hooks are lost, and rudiments of the

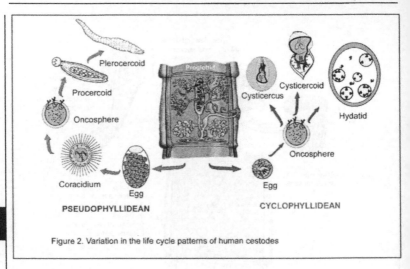

Figure 2. Variation in the life cycle patterns of human cestodes

Figure 2. Variations in the life cycle patterns of cestodes.

holdfast develop on an adult-like scolex. Human infection occurs when the second intermediate host is ingested.

Cyclophyllidean Life Cycle

The tapeworms in this group produce eggs that are usually fully embryonated and infective when discharged from the definitive host, and only rarely is there an aquatic stage in the life cycle. The eggs do not hatch until eaten by an intermediate host. The oncosphere that hatches from the egg penetrates the host's gut and develops into one of several morphological types of larvae characteristic of different species of tapeworms (Fig. 2). The following types of larvae are distinguished:

1. Cysticercoid: A cysticercoid larva consists of an anterior vesicle containing the scolex, which is not invaginated, and a tail-like posterior region containing the embryonic hooks. This is the larval stage in the life cycles of *Hymenolepis nana*, *H. diminuta* and *Dipylidium caninum*. This stage is usually found in invertebrate intermediate hosts. The term cercocystis is usually applied to a cysticercoid with a tail-like appendage from the bladder.

2. Cysticercus: A cysticercus is a larval stage characterized by a fluid-filled bladder surrounding a single scolex, retracted and invaginated within itself. This is a stage in the life cycles of some species of *Taenia*. The larval stage is usually found in vertebrates.

3. Coenurus: This larval stage is comprised of groups of scolices (protoscolices) which are produced by budding directly from the inner wall of a large bladder (occasionally called a cyst). The protoscolices, however, remain connected by a stalk to the wall of the bladder. This larval stage is characteristic of *Taenia multiceps*, and is found in vertebrates.

4. Hydatid: This larval stage is comprised of many (up to several million) protoscolices which usually develop by budding from the inner wall of vesicles of "brood capsules". Several liters of fluid may be produced within a hydatid. This larval form is known only for the members of the genus *Echinococcus* (Fig. 2).

Pseudophyllideans

Diphyllobothrium latum is the broad or fish tapeworm of humans. The parasite has a worldwide distribution, but a high incidence of infection occurs in Scandinavia, Finland, Alaska, and around the Great Lakes in both the United States and Canada. Other endemic areas include central Africa, parts of Asia, northern Chile and Argentina, and in New South Wales, Australia. Definitive hosts include humans, dogs, cats, pigs, and other fish-eating mammals.

The adult tapeworm lives primarily in the ileum, attached to the intestinal mucosa by an elliptical scolex bearing two bothria. The worm may be ten or more meters in length, and possess up to 4000 proglottids. Usually only one tapeworm is found per host. The life cycle requires two intermediate hosts. Unembryonated eggs, measuring about 40-50 μm wide by 60-70 μm long are continuously discharged from the gravid proglottids through the uterine pores and pass in the feces. Embryonation occurs in fresh-water and, in one to two weeks, the ciliated coracidium hatches through the operculum and swims about. In order for development to proceed, the coracidium must be ingested (within 12 to 24 hours) by the first intermediate host, copepods of the genera *Diaptomus* or *Cyclops*, in which development of the procercoid is completed (Fig. 3). The procercoid averages about 500 μm in length. When an infected copepod is ingested by freshwater fish, including pike, perch, trout, and salmon, the liberated procercoid burrows through the intestinal wall and into the flesh of the fish, where it develops into a plerocercoid. Mature plerocercoids vary in length from a few millimeters to several centimeters. Infection of humans occurs when fish harboring plerocercoids are consumed. Upon their release in the intestine of the definitive host, the parasites attach to the intestinal wall and develop to sexual maturity in 5-6 weeks.

Except for some nonspecific abdominal symptoms, most infections are symptomatic. Clinical B12 deprivation occasionally develops as a result of attachment of the parasite to the proximal portion of the jejunum. Quinicrine hydrochloride and niclosamide have been used to treat diphyllobothriasis.

Two other species, *D. chordatum* and *D. pacificum*, parasitize sea lions in northern and southern hemispheres and are apparently acquired by humans from eating raw marine fish. *Diphyllobothrium ursi* of bears also occurs in humans in Alaska. *Diplogonoporus grandis*, a parasite of whales, is frequently encountered in humans in Japan. Infection is probably acquired from eating raw anchovies, sardines, or other marine fish containing plerocercoids.

Sparganosis

Sparganosis is a disease of humans caused by migrating plerocercoids or spargana of certain diphyllobothrid tapeworms that, as adults, normally parasitize other mammals. Several species of *Spirometra* are intestinal tapeworms of feline and canine hosts. The first intermediate host is Cyclops, and various vertebrates such as frogs,

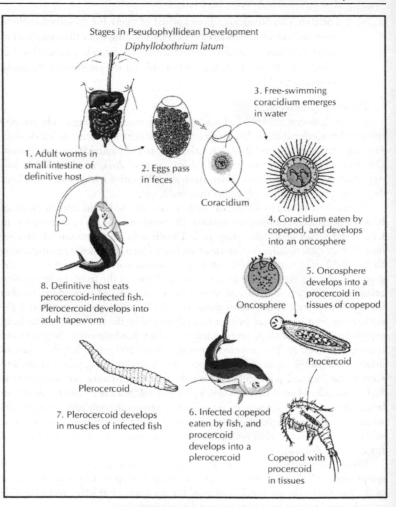

Stages in Pseudophyllidean Development
Diphyllobothrium latum

1. Adult worms in small intestine of definitive host

2. Eggs pass in feces

Coracidium

3. Free-swimming coracidium emerges in water

4. Coracidium eaten by copepod, and develops into an oncosphere

Oncosphere

5. Oncosphere develops into a procercoid in tissues of copepod

Procercoid

8. Definitive host eats perocercoid-infected fish. Plerocercoid develops into adult tapeworm

Plerocercoid

7. Plerocercoid develops in muscles of infected fish

6. Infected copepod eaten by fish, and procercoid develops into a plerocercoid

Copepod with procercoid in tissues

Figure 3. Life cycle of the pseudophyllidean *Diphyllobothrium latum*, the fish tapeworm of humans.

snakes, birds, pigs, and other mammals normally serve as second intermediate hosts by harboring the plerocercoids. Human infection is usually acquired by drinking water containing procercoid-infected copepods. The released procercoids migrate through the gut and develop into sparganum in muscles or subcutaneous tissues. Infection may result also from the consumption of inadequately cooked tissues of second intermediate hosts. The viable plerocercoids present in these hosts may become established in humans. In many regions of the Orient, snakes and tadpoles traditionally are consumed raw as a therapeutic measure for various illnesses. Infection may be acquired from the Oriental practice of poulticing open wounds, inflamed eyes or vagina, and numerous other lesions, with frog or snake flesh incidentally contaminated with plerocercoids. The larva penetrates the poulticed lesion and becomes established.

Spargana may persist in human tissues for several years and attain a length up to several centimeters. During development, they may produce severe inflammatory responses in surrounding tissues. In China and Vietnam, ocular sparganosis is a very serious infection, producing intense pain, periorbital edema, excessive lacrimation, nodule formation and corneal ulceration. Treatment of most cases of sparganosis is by surgical removal of the parasite (Fig. 4). A few cases are treated by chemotherapy.

Cyclophyllidean Life Cycles

Taenia saginata, the beef tapeworm, is the most common taeniid of humans. The parasite has a cosmopolitan distribution, occurring where inadequately cooked beef is eaten and also where sanitation is given little attention. Adult worms are up to 25 meters long, but worms slightly less than one-third this size are more common. The scolex bears four muscular suckers. The terminal gravid proglottids, each of which contains about 100,000 eggs, detach from the strobila in the intestine of humans and pass out with the feces. Some proglottids exhibit marked individual activity, migrating out through the anus and onto the skin in the perianal region. At this stage, they may be mistaken for adult trematodes. When the evacuated proglottids rupture, fully embryonated eggs are discharged. The spherical eggs remain infective for several weeks. Various ruminants serve as intermediate hosts, including cattle, goat, bison, sheep, giraffe, and llamas.

Following ingestion of the eggs by a ruminant, the outer shell disintegrates in the duodenum releasing the six-hooked oncosphere. Histologic secretions from the oncosphere facilitate its penetration through the mucosa and into the circulation. The parasite enters muscles in various regions of the body and develops into an infective cysticercus in about 2 months. *Cysticerci*, which measure up to 10 mm long by 6 mm wide, remain viable up to 9 months. Eventually, the cysticerci are destroyed by what appears to be a host response involving the development of a fibrous capsule around the parasites. Prior to being identified as a larval stage, the cysticercus of *T. saginata* was given a separate species status and referred to as *Cysticercus bovis*. The disease in cattle was known as cysticercosis bovis.

Human infection is acquired through the ingestion of raw or inadequately cooked beef containing cysticerci (Fig. 5). A period of about three months is required for the ingested cysticerci to develop into sexually mature worms in the intestine. Although autoinfection is possible if eggs voided in the feces are swallowed, human cysticercosis, involving *T. saginata*, is rarely encountered. Cattle and other ruminants become infected when their feed and feeding areas are contaminated with the egg-laden feces of infected humans. The indiscriminate release of raw sewage into lakes and rivers contributes significantly to the dissemination of the eggs of the parasite.

Taeniasis caused by *T. saginata* may be characterized by nausea, persistent abdominal pain, localized sensitivity to touch, diarrhea, or alternating diarrhea and constipation, loss of appetite and weight, chronic indigestion, and hunger pains. Intestinal obstruction is rarely observed. Psychological stress resulting from gravid proglottids migrating out of the anus of an infected individual may be severe.

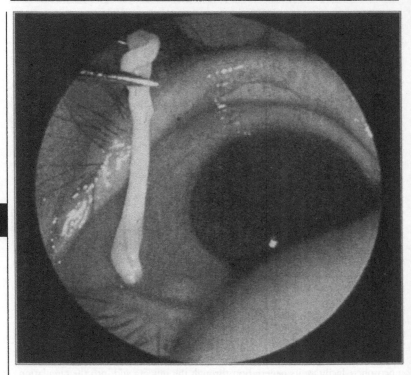

Figure 4. Ocular sparganosis. FromYamaguchi, T. (Ed.) 1981. A Color Atlas of Clinical Parasitology. (Wolfe Medical Publ. Ltd.) Lea & Febiger, Pennsylvania.

The pork tapeworm, *Taenia solium*, has a worldwide distribution. Adult worms, which are parasitic in humans only, are up to 7 meters long. The scolex bears four suckers and a double crown of prominent rostellar hooks. Gravid proglottids detach from the strobila and are evacuated with the feces. Each proglottid contains about 30,000 eggs. The eggs, which are indistinguishable from those of *T. saginata*, are fully embryonated when discharged from the host's intestine (Fig. 6). The life cycle is similar to that of *T. saginata* except that pigs are the intermediate hosts harboring the cysticerci (*Cysticercus cellulosae*) (Fig. 5).

Taenia solium is potentially the most serious tapeworm of humans, primarily because individuals harboring the adult worm may also develop cysticercosis (Fig. 7) through autoinfection. Unlike *T. saginata*, the cysticerci of *T. solium* develop readily in various tissues of humans, which then become potential intermediate hosts (Fig. 8). Both internal and external autoinfection are possible. External autoinfection results when embryonated eggs are ingested. This frequently results from the fecal contamination of food and drink. Internal autoinfection occurs when detached proglottids are carried by reverse peristalsis from the small intestine into the stomach. Here the released eggs hatch and the oncospheres migrate into the body and undergo development to become cysticerci. Light infections may be asymptomatic.

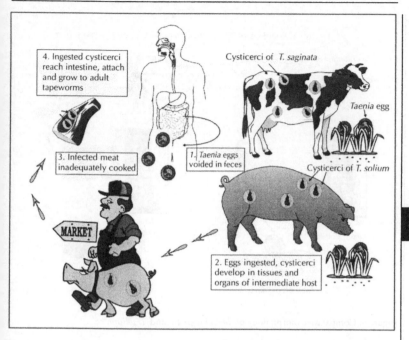

Figure 5. Life cycles of the beef tapeworm, *Taenia saginata*, and the pork tapeworm, *T. solium*.

Apparently, no tissue or organ is resistant to penetration by cysticerci. In heavy infections, the larvae may develop in the brain, heart or eyes, causing numerous serious problems such as paralysis, epileptic attacks, disequilibrium, blindness, and hydrocephalus as a result of blockage and inadequate drainage of cerebrospinal fluid. Ocular cysticercosis may be mistakenly diagnosed as a malignancy, resulting in the unwarranted removal of the eye. In various tissues and organs, host cellular reactions occur, and as a result the cysticerci become calcified and/or enveloped by a fibrous capsule. Cysticerci may remain viable for a few years before they begin to degenerate, whereupon an intense local inflammatory response is initiated by the host.

Adults of *Taenia (Multiceps) multiceps* parasitize carnivores, particularly dogs and other canines in many parts of the world. The larval stage, which is similar in morphology to a hydatid and is termed a coenurus, occurs in herbivorous mammals. In sheep, the most common intermediate host, coenuriasis is characterized by a vertigo resulting in an unstable gait or giddiness, hence the designation "gid" for the infection. Occasionally, coenuri develop in the subcutaneous tissues, muscles, brain, spinal cord, and eyes of humans. These infections result from the accidental ingestion of eggs. The oncosphere released from the egg penetrates the intestinal mucosa and enters the general circulation. Multiple scolices then bud from the inner wall of each parasite to form a coenurus. Unfortunately, no effective therapeutic procedure is available.

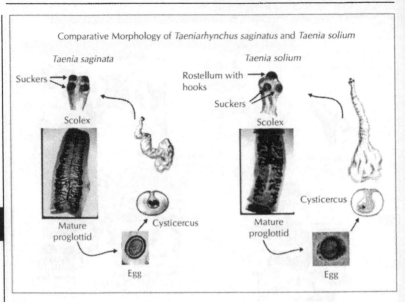

Comparative Morphology of *Taeniarhynchus saginatus* and *Taenia solium*

Figure 6. Comparative morphology of *Taenia saginata* and *T. solium*.

Echinococcus *Species and Hydatidosis*

The adult tapeworms of the genus *Echinococcus* are the smallest taeniids. The entire worm is usually less than a centimeter long and is composed of a scolex with an armed rostellum, a neck region, and 3 or 4 proglottids. The adult worms live attached to the intestinal mucosa of various carnivores, especially dogs and other canines, and felines, but not in humans (Fig. 9). The larval stage, or hydatid, is the causative agent of hydatidosis, a disease of considerable medical and veterinary importance. Hydatids occur in virtually any mammal that ingests the eggs of *Echinococcus*. When swallowed by the intermediate host, the eggs hatch and the oncospheres penetrate the gut wall and enter the venules of the hepatic portal veins. Hydatids may develop in any organ or tissue, but commonly are found in the liver, lungs, marrow cavity of long bones, kidneys, spleen, muscles, and brain (Fig. 10). Growing hydatids occupy space, exert pressure, and destroy surrounding host tissues. From the inner germinative surface of the hydatid there develop secondary vesicles or brood capsules, within which the larval stages with immature, inverted scolices (e.g., protoscolices) develop. Occasionally, the parent hydatid cyst forms 'daughter cysts', which then produce brood capsules. Hydatid cysts commonly range in size from 5 to 10 cm and contain several liters of fluid, but cysts up to 50 cm and containing up to 18 liters of fluid have been found. The reproductive potential of the organism is immense, with each liter of fluid containing up to one million protoscolices. Eventually, the brood capsules and daughter cysts break down, and the liberated components, termed hydatid sand, float free in the hydatid fluid. When infected viscera are ingested by canine definitive hosts, the cyst wall is digested,

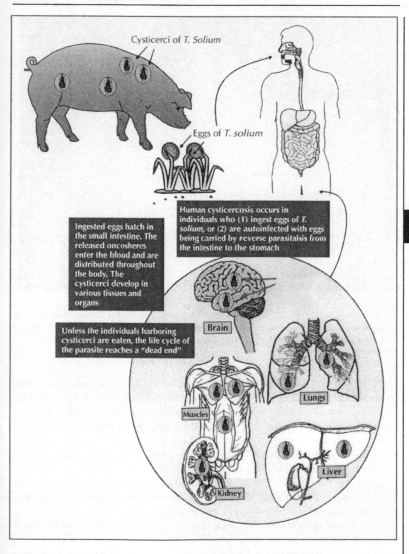

Cysticerci of *T. Solium*

Eggs of *T. solium*

Ingested eggs hatch in the small intestine. The released oncosheres enter the blood and are distributed throughout the body. The cysticerci develop in various tissues and organs

Human cysticercosis occurs in individuals who (1) ingest eggs of *T. solium*, or (2) are autoinfected with eggs being carried by reverse parasitalsis from the intestine to the stomach

Unless the individuals harboring cysticerci are eaten, the life cycle of the parasite reaches a "dead end"

Brain

Lungs

Muscles

Kidney

Liver

6

Figure 7. Development of human cysticercosis.

liberating the protoscolices, which then evaginate, attach to the intestinal wall, and develop into adult parasites.

Three species of *Echinococcus* are known to cause hydatidosis in humans. *Echinococcus granulosis* matures in dogs and other canines (but not foxes) and uses mainly herbivores as intermediate hosts. The parasite is cosmopolitan in distribution, but it is especially common in areas where domestic herbivores, such as sheep, pigs, goats, horses, rabbits, camels, and reindeer, are raised in association with dogs. Dogs ac-

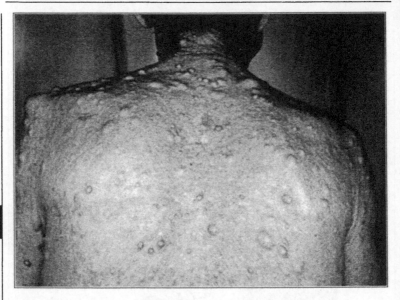

Figure 8. Human cysticercosis. This individual harbors the larval stage of *Taenia solium*, and thus serves as a potential intermediate host for the tapeworm. FromYamaguchi, T. (Ed.) 1981. A Color Atlas of Clinical Parasitology. (Wolfe Medical Publ. Ltd.) Lea & Febiger, Pennsylvania.

quire the infection when they consume the offal of butchered animals, while herbivores become infected when they feed on vegetation contaminated with egg-laden dog feces. Human infection occurs when *Echinococcus* eggs are ingested, usually as a result of fondling infected dogs or handling their scats. A second species, *E. multilocularis*, matures mainly in foxes, but cats, dogs, and other canines also serve as definitive hosts. Small rodents, such as mice, voles, and lemmings, are intermediate hosts. The parasite has been reported from parts of Europe, North and South America, and New Zealand. A third species, *E. oligarthrus*, is a parasite of jaguars, pumas, and other wild felines. Human infections with the hydatid of this parasite have been reported from South and Central America.

The symptoms of hydatidosis depend on the type and location of the hydatid. Hepatic hydatidosis produces jaundice, ascites, and splenomegaly, which frequently results in fatal intrahepatic portal hypertension. Osseous hydatids progressively erode bone causing fractures and eventual decay. The symptoms of cranial hydatids mimic those of intracranial tumors. Chemotherapy is ineffective in hydatid disease. Surgical removal of the cyst is the only successful method of treatment, providing the hydatids are situated in operable locations. Considerable care is required during surgery to prevent the hydatid fluid from escaping into the body cavity, as this may cause secondary infections and fatal anaphylactic shock. Moreover, scolices liberated from ruptured hydatids may become attached to the peritoneum and produce multiple secondary growths. In some cases, it may be possible to first aspirate the fluid with a syringe before excising the hydatids.

Life Cycle of *Echinococcus*

2. Eggs passed in feces are ingested by various intermediate hosts

1. Adult worms in small intestine

4. Definitive hosts become infected when they ingest intermediate hosts containing hydatids

3. Eggs ingested by intermediate hosts hatch in small intestine. The freed oncospheres dispense to various organs and develop into hydatids

Hydatid

Human hydatidosis results from ingesting eggs of Echinococcus. The disease occurs from intimate contact with dogs. In humas the life cycle is a dead end.

Figure 9. Life cycle of *Echinococcus*.

Dipylidium caninum is a cosmopolitan tapeworm of dogs and cats, measuring up to 50 cm in length. The scolex possesses four conspicuous suckers and a retractable rostellum with several rows of minute hooklets. Detached gravid proglottids are passed with the feces or actively migrate out the anus. On reaching the soil, the proglottids begin to desiccate, releasing egg capsules, each of which contains several embryonated eggs. Fleas, chewing lice, and possibly other arthropods serve as intermediate hosts when they ingest egg capsules (Fig. 11). The parasite invades the host's hemocoel and develops into a cysticercoid. The definitive host becomes in-

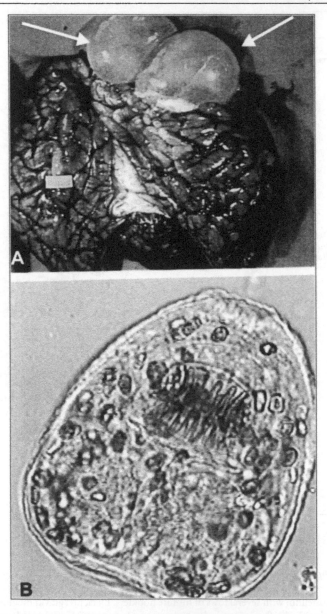

Figure 10. (A) Two fluid filled hydatid cysts (arrows) situated on the surface of the brain. From Yamaguchi, T. (Ed.) 1981. A Color Atlas of Clinical Parasitology. (Wolfe Medical Publ. Ltd.) Lea & Febiger, Pennsylvania. (B) Photograph of a hydatid cyst showing numerous protoscolices that can form daughter cysts. Rupture of the cysts can cause secondary hydatidosis. Courtesy of Dr. Pietro Caramello, and the Carlo Denegri Foundation, Torino, Italy.

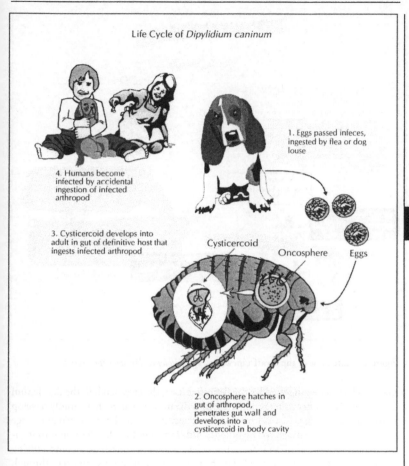

Life Cycle of *Dipylidium caninum*

4. Humans become infected by accidental ingestion of infected arthropod

1. Eggs passed in feces, ingested by flea or dog louse

3. Cysticercoid develops into adult in gut of definitive host that ingests infected arthropod

Cysticercoid   Oncosphere   Eggs

2. Oncosphere hatches in gut of arthropod, penetrates gut wall and develops into a cysticercoid in body cavity

Figure 11. Life cycle of the dog tapeworm, *Dipylidium caninum*.

fected when infected arthropod hosts are consumed. The cysticercoids escape in the intestine and develop directly to adult worms in about one month. Human infections with *D. caninum* frequently involve children who fondle infected animals and accidentally ingest infected fleas.

Hymenolepis (Vampirolepis) nana, the dwarf tapeworm, is a cosmopolitan parasite of rodents and other mammals including humans. The parasite is the most common human tapeworm in the world, occurring most frequently in young children. The parasite has a worldwide incidence of about 1 percent. The adult tapeworm ranges from 0.1 to 10 cm in length. In heavy infections, where crowding affects size, the worms are small, averaging less than 3 cm. The scolex bears four deep acetabula and a retractable rostellum armed with a crown of hooks. The eggs are usually liberated from the gravid proglottids before they detach from the strobila. Eggs are infective to the definitive host, in which both larval and adult stages

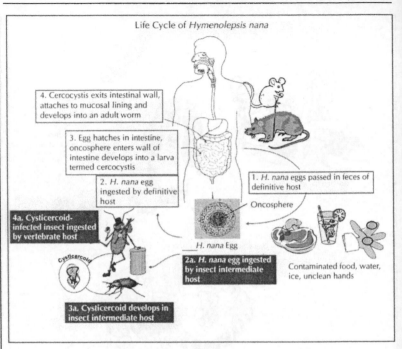

Figure 12. Life cycle of the dwarf tapeworm, *Vampirolepis (Hymenolepis) nana*.

may develop. When ingested by the definitive host, the eggs hatch in the duodenum and the liberated oncospheres penetrate into the mucosa where they rapidly develop into cysticercoids (Fig. 12). Within one-week post-infection, the cysticercoids emerge into the gut lumen, attach to the mucosa and, in about 2 weeks, develop into sexually mature worms. Human infection is believed to be acquired most commonly by the ingestion of eggs that are voided in the feces. Heavy infections can occur through internal autoinfection due to the hatching of eggs released from adult worms in the small intestine. The life cycle of *H. nana* is of interest in that the parasite may develop to the cysticercoid stage in intermediate hosts, which include a large number of insects (e.g., adult grain or flour beetles, cockroaches, fleas). When these intermediate hosts are consumed by humans (or other suitable definitive hosts), the adult tapeworms develop in the small intestine. Light infections are usually asymptomatic. Abdominal pain, diarrhea, anorexia, vomiting, pruritus of the nose and face and urticaria generally characterize heavy infections.

Hymenolepis diminuta is a parasite of rats and occasionally humans. It is a larger species than *V. nana*, measuring between 20 and 90 cm in length. The scolex possesses the typical tetrad of suckers, but lacks an armed rostellum. Completion of the life cycle requires an arthropod intermediate host, commonly flour moths, stored grain beetles, earwigs, and rodent fleas. The intermediate hosts, which harbor the cysticercoids, become infected from eating flour or cereal foods contaminated by the egg-laden dropping of infected rats. Human infections are associated primarily with the contamination of cereals and other food by infected grain beetles. Infections are generally mild, with occasional indigestion and slight abdominal pain. In some humans, the adult worms are spontaneously evacuated, suggesting an unsuitable host environment.

CHAPTER 7

General Morphology of Parasitic Nematodes

Nematodes are unsegmented worms that typically are elongate and cylindrical in shape with tapered ends. Roundworms vary considerably in size from microscopic to over a meter in length. They possess a fluid-filled cavity or pseudocoel, and a complete digestive system comprised of an anterior mouth, a muscular esophagus (pharynx), an intestine, and a rectum that terminates posteriorly at the anus. The mouth may be surrounded by lips and a buccal cavity with cutting plates or teeth. Parasitic nematodes commonly feed on the semi-liquid contents of the host's alimentary canal, intestinal mucosa, blood or other body fluids, and various lysed tissues. In some forms, the muscular esophagus is cylindrical and virtually of a uniform diameter throughout, and the parasite is termed filariform. This type of esophagus generally characterizes infective stage larvae. In certain other nematodes, the esophagus is expanded posteriorly into a valved bulb, and the parasite is termed rhabditiform (Fig. 1). The latter type frequently characterizes the free-living larval stages. Some nematodes have both filariform and rhabditiform stages in their life cycles. The body is covered by a non-cellular cuticle, which may exhibit longitudinal ridges, striations, wart-like structures, lateral expansions anteriorly and posteriorly, and spines. Before reaching sexual maturity, all nematode larvae undergo a series of four molts or ecdyses. Cuticular structures of some importance are amphids and phasmids. Amphids are a pair of minute sensory organs, considered to be chemoreceptors that open on each side of the head. Phasmids are a pair of caudal (post-anal) organs similar in structure to amphids. Some phasmids are glandular and serve an excretory function, while others are sensory and believed to be involved in chemoreception.

With few exceptions, nematodes are dioecious and exhibit sexual dimorphism, with males generally smaller and possessing a more curved tail than females. Monoecious species may be either parthenogenic or self-fertilizing hermaphrodites. The gonads are tubular cords of cells continuous with the ducts that transport the gametes to the external environment. In males, the intestine and the reproductive tract open into the cloaca. There is a single tubular testis from which extends a vas deferens that terminates in a musuclar ejaculatory duct. Most males have a pair of sclerotized copulatory spicules originating from within the cloacal wall that serve as holdfast structures during copulation. The tail may be drawn out into longitudinal ridges called caudal alae, which also assist in holding the female in opposition during copulation. The female reproductive system also is tubular and frequently highly coiled within the body. The system may be composed of a single set of reproductive structures, but commonly there are two sets, and occasionally more, with the following regions differentiated; ovary, oviduct, seminal receptacle, uterus, ovijector, and vagina. The vulva (genital pore) is located ventrally and is independent of the alimentary

Parasites of Medical Importance, by Anthony J. Nappi and Emily Vass.
©2002 Landes Bioscience.

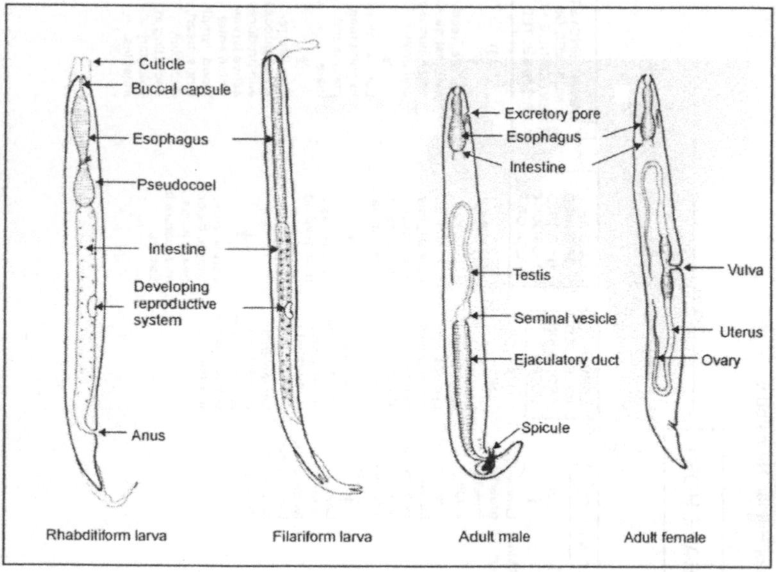

Figure 1. Generalized diagram of the larval and adult stages of Nematodes. Modified from Meyer, M. C. and Olsen, O. W. 1980. Essentials of Parasitology. W. C. Brown Co., Dubuque, Iowa.

canal. Experimental alterations in the environment have been shown to induce changes in the sexual characteristics of certain nematodes, including sex reversals.

There are two basic types of life cycles among parasitic nematodes, direct (monoxenous) with only a single host, or indirect (heteroxenous) with two or more hosts involved. Nematode larvae develop through a series of four stages, each separated by a molting of the cuticle. The third stage is generally infective to the final host. In some species one or more larval stages may occur within the egg capsule. In many parasitic species with a direct life cycle, the first three larval stages are free in the soil, where the first two stages feed on soil components. The third stage retains the shed cuticle of the preceding stages as an enclosing sheath and is unable to feed. Oviparous species produce eggs that are released from the definitive host with the feces. In ovoviviparous species, the eggs hatch *in utero* and the larvae are passed in the feces. Some of the variations in the life cycle patterns of parasitic nematodes are given in Table 1.

Trichuris trichiura

Trichuris trichiura, or whipworm, has a cosmopolitan distribution, but is more frequently encountered among the poor in the tropics and subtropics, and where living conditions are crowded and sanitation is poor. In parts of the southeastern United States the incidence of trichuriasis may reach 25%, with young children most commonly infected. The adult worms measure from 3 to 5 cm in length, with females larger than males. The anterior three-fifths of the body is thread-like or

Table 7.1 Some Variations in the Life Cycles of Human Nematodes

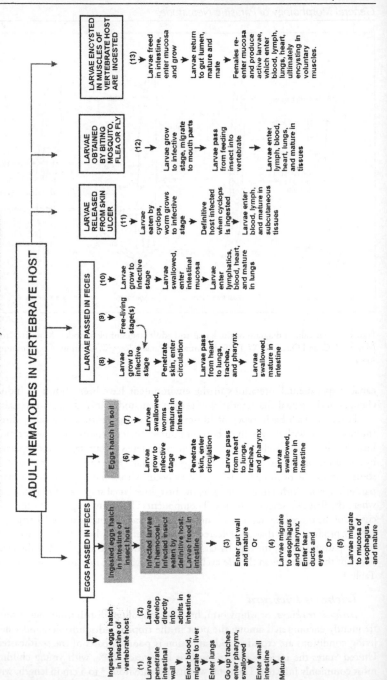

ADULT NEMATODES IN VERTEBRATE HOST

EGGS PASSED IN FECES

Ingested eggs hatch in intestine of vertebrate host

(1) Larvae penetrate intestinal wall

Enter blood, migrate to liver

Enter lungs

Go up trachea, enter pharynx, swallowed

Enter small intestine

Mature

(2) Larvae develop directly into adults in intestine

Ingested eggs hatch in intestine of insect host

Infected larvae in hemocoel. Infected insect eaten by definitive host. Larvae freed in intestine

(3) Enter gut wall and mature

Or

(4) Larvae migrate to esophagus and pharynx. Enter tear ducts and eyes

Or

(5) Larvae migrate to mucosa of esophagus, and mature

Eggs hatch in soil

(6) Larvae grow to infective stage

Penetrate skin, enter circulation

Larvae pass from heart to lungs, trachea, and pharynx

Larvae swallowed, mature in intestine

(7) Larvae swallowed, worms mature in intestine

LARVAE PASSED IN FECES

(8) Larvae grow to infective stage

Penetrate skin, enter circulation

Larvae pass from heart to lungs, trachea, and pharynx

Larvae swallowed, mature in intestine

(9) Free-living stage(s)

(10) Larvae grow to infective stage

Larvae swallowed, enter intestinal mucosa

Larvae enter lymphatics, blood, heart, and mature in lungs

LARVAE RELEASED FROM SKIN ULCER

(11) Larvae eaten by cyclops, worm grows to infective stage

Definitive host infected when cyclops is ingested

Larvae enter blood, lymph, and mature in subcutaneous tissues

LARVAE OBTAINED BY BITING MOSQUITO, FLEA OR FLY

(12) Larvae grow to infective stage, migrate to mouth parts

Larvae pass from feeding insect into vertebrate

Larvae enter lymph, blood, heart, and mature in tissues

LARVAE ENCYSTED IN MUSCLES OF VERTEBRATE HOST ARE INGESTED

(13) Larvae freed in intestine, enter mucosa and grow

Larvae return to gut lumen, mature and mate

Females re-enter mucosa and produce active larvae, which enter blood, lymph, lungs, heart, ultimately encysting in voluntary muscles.

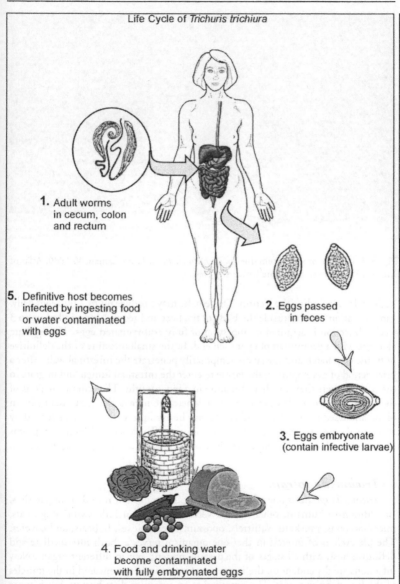

Life Cycle of *Trichuris trichiura*

1. Adult worms in cecum, colon and rectum

5. Definitive host becomes infected by ingesting food or water contaminated with eggs

2. Eggs passed in feces

3. Eggs embryonate (contain infective larvae)

4. Food and drinking water become contaminated with fully embryonated eggs

Figure 2. Life cycle of the whipworm, *Trichuris trichura*.

filariform, with the remaining portion abruptly expanded, resembling a lash or whip with a thick handle.

Adult whipworms may live several years embedded primarily in the mucosa of the cecum (Fig. 2). In heavy infections, however, the worms can be found elsewhere

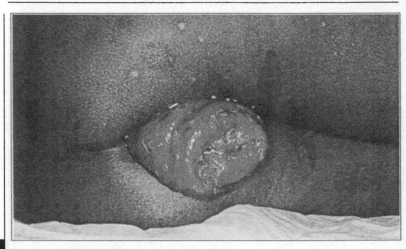

7

Figure 3. Prolapse of the rectum due to *Trichuris trichura*. From Zaman, V. 1980. Atlas of Medical Parasitology. Singapore University Press.

in the colon, including the rectum. The females may produce up to 7,000 eggs daily. Embryonation occurs outside the body of the host and requires a period of up to 3 weeks. Infection is acquired by ingestion of fully embryonated eggs. The eggs are elongate with a prominent plug at each end. In the small intestine of the definitive host, the eggs hatch and the larvae temporarily penetrate the intestinal wall. After a brief period of development, the larvae re-enter the intestinal lumen and migrate to the cecum where they attach and mature in a few months. The anterior regions of the worms are embedded in the mucosa, where the nematodes feed essentially on blood and lysed tissues. Light infections may be asymptomatic, or characterized by abdominal pain, diarrhea, constipation, vomiting, flatulence and fever. Symptoms of heavy infections may include bloody diarrhea, colitis, and prolapse of the rectum (Fig. 3).

Trichinella spiralis

Trichinella spiralis is one of the more clinically important roundworm parasites. In addition to humans, other natural vertebrate hosts include swine, dogs, cats, mice, raccoons, muskrats, squirrels, opossums, bears, foxes, bobcats, and coyotes. The life cycle is of interest in that one organism serves as both intermediate and definitive host, with all stages of the parasite present, but in different organs. New infections are dependent on the ingestion of infective larvae encased in the muscles of the first definitive host. The parasite has an extensive distribution, but is more common in areas where garbage containing uncooked pork scraps is fed to swine. It is estimated that 35 million persons in the United States are infected with trichinosis.

Adult *T. spiralis* is thread-like in shape and barely visible to the unaided eye. Females measure 3 to 5 mm in length, and males average slightly less than 2 mm in length. The parasites have a life span of a few weeks to three months. Human infection occurs by the ingestion of raw or insufficiently cooked meat, chiefly pork, con-

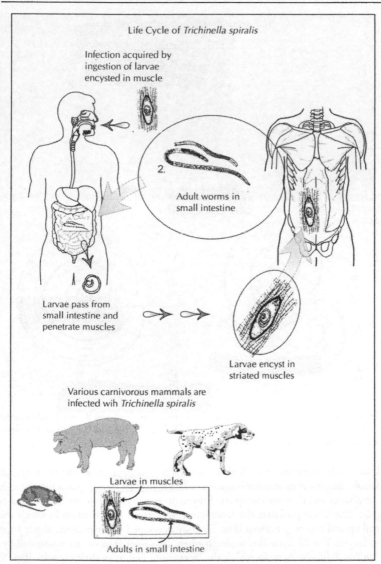

Life Cycle of *Trichinella spiralis*

Infection acquired by ingestion of larvae encysted in muscle

2.

Adult worms in small intestine

Larvae pass from small intestine and penetrate muscles

Larvae encyst in striated muscles

Various carnivorous mammals are infected wih *Trichinella spiralis*

Larvae in muscles

Adults in small intestine

Figure 4. Life cycle of *Trichinella spiralis*.

taining the infective larvae encysted in the muscle. Occasionally, infections result from ingestion of bear or walrus meat, and beef contaminated by a meat grinder previously used to grind infected pork. The larvae excyst in the small intestine and become attached to the mucosa and grow to the adult stage in less than two days, at which time mating takes place (Fig. 4). The intestinal phase of the infection generally

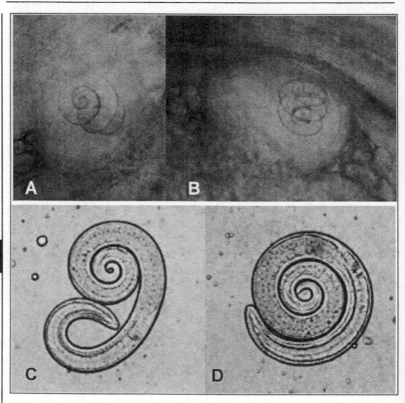

Figure 5. Larvae of *Trichinella* coiled and encysted in muscle (A and B), and freed from their cysts (C and D). Courtesy of the Centers for Disease Control and Prevention, Division of Parasitic Diseases, Atlanta, Georgia.

lasts less than two months. Within three days after copulation, the ovoviviparous females burrow deep into the intestinal wall and deposit larvae, which first enter the lymphatics and then are transported through the general circulation to various organs. The larvae penetrate the sarcolemma of skeletal muscle fibers in which they coil up and become encysted (Fig. 5). Fully developed larvae measure about 1 mm in length. The life cycle is completed when the encysted larvae are consumed by a suitable definitive host, in which the larvae are freed from their capsules (Fig. 5). Some of the encysted larvae may remain viable and dormant for many years, even after the capsules become calcified, which usually occurs 6-18 months post-infection. The muscles most commonly infected include the diaphragm, intercostals, abdominal, pectoral, gastrocnemius, deltoid, larynx, and base of the tongue.

Manifestations of trichinosis vary from asymptomatic cases to fatal infections. Fever and gastrointestinal complaints, particularly diarrhea, may be common during the early intestinal phase. During muscle invasion and encystment there may be severe myositis producing difficulty in breathing, mastication and speech, hemor-

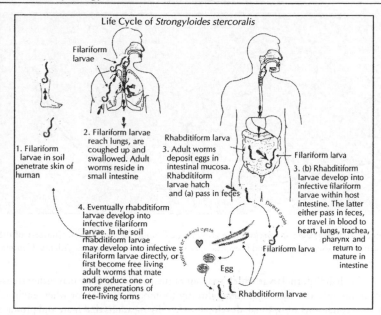

Figure 6. Life cycle of *Strongyloides stercoralis*.

rhages of the skin, mucous membranes, lungs, circumorbital edema, vasculitis, and neurological manifestations. On rare occasions, the larvae may penetrate the cardiac muscles, and pass through the cerebral capillaries causing motor and psychic disturbances.

Strongyloides stercoralis

Strongyloides stercoralis is an intestinal parasite of humans, dogs, cats, and other mammals. The parasite is commonly found in tropical and subtropical areas and is also sporadically reported from temperate and cold climates. A second species, *S. fuelleborni,* occurs commonly in central and east Africa. Both parasitic and free-living generations are formed. Parasitic females are protandrogonous, their reproductive organs developing after the male reproductive organs have disappeared. These females, which measure up to 2.5 mm in length, burrow into the intestinal mucosa where they lay partially embryonated eggs. Occasionally, the parasites are found in the bronchial passages, and biliary and pancreatic systems. The eggs hatch within the submucosa or during passage through the lumen of the intestine, liberating rhabditiform larvae, which are then voided with the feces. However, if delayed in their movement down the digestive tract, the rhabditiform larvae may molt and transform into filariform larvae, which penetrate the colonic mucosa or perianal skin, migrate via the blood to the lungs and eventually to the small intestine where they mature to the adult stage. This developmental cycle, termed autoinfection, may account for some infections persisting for several years.

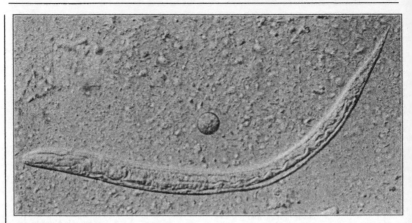

Figure 7. *Strongyloides stercoralis.* Note also cyst of *Entamoeba coli* (arrow). Courtesy of the Centers for Disease Control and Prevention, Division of Parasitic Diseases, Atlanta, Georgia.

The rhabditiform larvae which pass out of the host in the feces may either transform into infective filariform larvae directly (homogonic cycle), or when environmental conditions are optimal, develop into free-living adults that mate and produce eggs that hatch into rhabditiform larvae (Figs. 6 and 7). This indirect or sexual cycle (heterogonic cycle) may be repeated indefinitely before infective filariform larvae are produced. In the soil, the filariform larvae are unable to develop further. The life cycle continues when the filariform larvae enter a host, either by skin penetration or ingestion. If infection occurs by penetration of the skin of the host, the larvae enter the blood and are transported to the lungs where they pass from the capillaries into the alveoli, up the bronchial passages to the pharynx where they are coughed up, swallowed, and later complete their maturation in the small intestine. A lung migration apparently does not occur when infection is acquired by ingestion of the filariform larvae.

Slight hemorrhage, swelling and itching are sometimes noted at the sight of penetration of the skin by the invasive filariform larvae. Other symptoms of strongyloidiasis include pneumonitis, anemia, abdominal pain, ulceration and sloughing of the intestinal mucosa, and occasionally death resulting from septicemia. Intestinal malabsorption syndrome with steatorrhea is sometimes observed. Thiabendazole is currently the drug used in treatment of strongyloidiasis.

Hookworms

Several species of hookworm infect domestic and wild animals. Two species, *Ancylostoma duodenale* and *Necator americanus*, are parasitic in humans. Both species are found around the world, but *A. duodenale* or Old World hookworm occurs mainly in Europe, Asia and Africa, and is found only in scattered areas of the Caribbean Islands, South America and the United States. *Necator americanus*, the New World or American hookworm is the predominant species in the United States. In *Ancylostoma*, the buccal capsule contains paired tooth-like processes (Fig. 8), while in *Necator* the buccal capsule is provided with semilunar cutting plates. Adult males

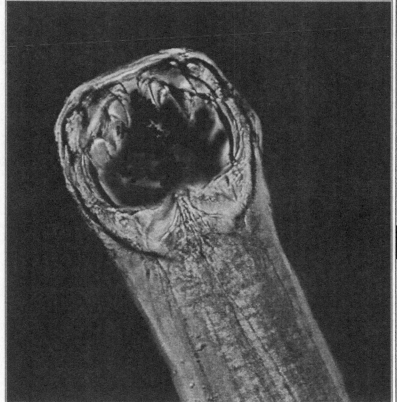

Figure 8. Buccal capsule of the hookworm *Ancylostoma caninum* illustrating three pairs of cutting teeth. From Zaman, V. 1980. Atlas of Medical Parasitology. Singapore University Press.

are 7 to 11 mm long, and females 8 to 13 mm long. The adults live in the small intestine attached to the mucosa, feeding on blood and tissue fluids. The eggs are passed in the feces and under optimal conditions of moisture and temperature, rhabditiform larvae hatch within 24 hours. The rhabditiform larvae feed on fecal matter, bacteria, and organic material in the soil. Eventually, they transform into third-stage, non-feeding, filariform or infective larvae. The filariform larvae penetrate the skin of the host, and enter the circulatory system. The larvae are carried by the blood to the heart and into the lungs (Fig. 9). They emerge from the capillaries into the alveolar spaces, move up the bronchial passages to the pharynx and are swallowed. The worms attach to the intestinal mucosa where they molt twice to become sexually mature. The worms mate and produce eggs that are passed in the feces. Three to six weeks are required from the time of infection to the appearance of eggs in the feces. In some cases, transplacental transmission and transmission in breast milk are known to occur.

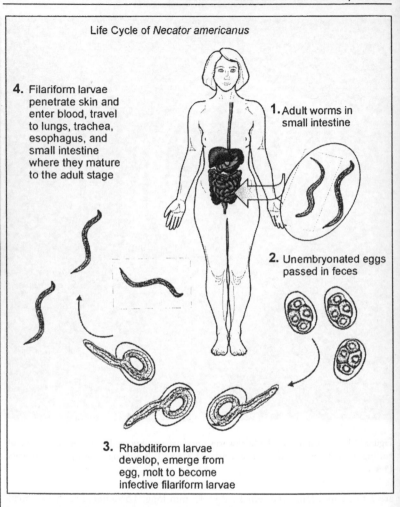

Life Cycle of *Necator americanus*

4. Filariform larvae penetrate skin and enter blood, travel to lungs, trachea, esophagus, and small intestine where they mature to the adult stage

1. Adult worms in small intestine

2. Unembryonated eggs passed in feces

3. Rhabditiform larvae develop, emerge from egg, molt to become infective filariform larvae

Figure 9. Life cycle of the hookworm, *Necator americanus.*

Hookworms are among the more prevalent parasites of humans. Clinical manifestations of the disease are dependent on the number of worms present, the age and nutritional state of the infected individual, and the immune capabilities of the host. Penetration of the skin by the filariform larvae may cause an allergic or urticarial reaction known as ground itch. Secondary infection with pyogenic bacteria may occur. During larval migration through the lungs, patients may experience fever, headache, nausea, dyspnea, excessive coughing and pharyngeal soreness. Pneumonitis may occur in severe infections as a result of the parasites rupturing through the capillaries and invading the alveoli. During the intestinal phase of the infection, the symptoms may include such nonspecific gastrointestinal disorders as intermittent

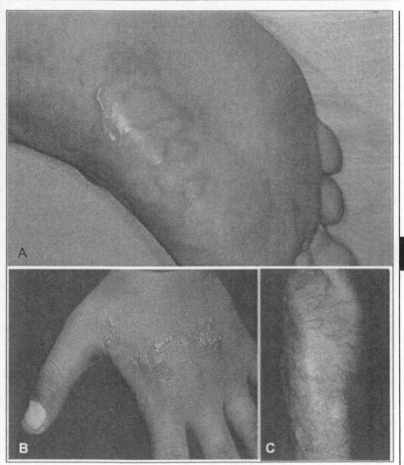

Figure 10. Cutaneous larva migrans infection ("creeping eruption") by cat or dog tapeworms. (A) Courtesy of Dr. Pietro Caramello, and the Carlo Denegri Foundation, Torino, Italy. (B) From Zaman, V. 1980. Atlas of Medical Parasitology. Singapore University Press. (C) From Yamaguchi, T. (Ed.) 1981. A Color Atlas of Clinical Parasitology. (Wolfe Medical Publ. Ltd.) Lea & Febiger, Pennsylvania.

abdominal pains, nausea, loss of appetite, vomiting, flatulence, diarrhea and/or constipation. Intestinal malabsorption is not common with hookworm infection. Major pathological manifestations of hookworm infection are iron-deficiency anemia, protein deficiency, and immunodeficiency. Since as much as 200 ml of blood per day may be lost by patients with heavy hookworm infections, a large parasite burden can gradually deplete a patient's serum proteins and iron reserves. Chronic malnutrition, especially in young patients, compounds the problems of hookworm infection, and may adversely affect physical and mental development.

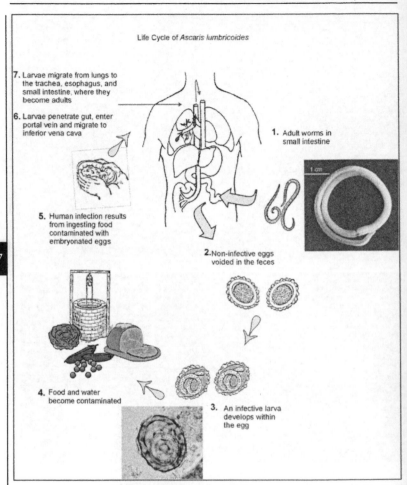

Life Cycle of *Ascaris lumbricoides*

7. Larvae migrate from lungs to the trachea, esophagus, and small intestine, where they become adults

6. Larvae penetrate gut, enter portal vein and migrate to inferior vena cava

1. Adult worms in small intestine

5. Human infection results from ingesting food contaminated with embryonated eggs

2. Non-infective eggs voided in the feces

4. Food and water become contaminated

3. An infective larva develops within the egg

Figure 11. Life cycle of *Ascaris lumbricoides*.

Drugs for the treatment of hookworm infection include bephenium hydroxynaphthoate (Alcopara), pyrantel pamoate, and tetrachloroethylene. The only definitive diagnosis of hookworm infection is the early examination of feces and the identification of parasite eggs. Old stools or feces from constipated patients may contain, in addition to eggs, hatched rhabditiform larvae.

Cutaneous Larva Migrans

Two common hookworms of domestic dogs and cats, *Ancylostoma braziliense* and *A. caninum*, have been found in humans on several occasions. Two other hookworm species that have been found in humans are *A. ceylanicum*, which normally parasitizes carnivores in parts of Asia and the East Indies, and *A. malayanum*, which

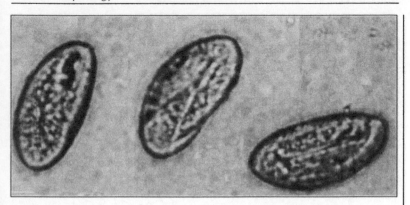

Figure 12. Eggs of *Enterobius vermicularis*. Sizes range from 20-30 mm by 50-60 mm. Courtesy of the Centers for Disease Control and Prevention, Division of Parasitic Diseases, Atlanta, Georgia.

is parasitic in bears in Malaysia. When nonhuman hookworm filariform larvae penetrate human skin, they may be trapped within the integument, unable to complete their migratory cycle to the lungs and intestinal mucosa. Instead, they migrate randomly through the subcutaneous tissue forming serpiginous burrows and causing intense itching, a condition referred to as cutaneous larva migrans or creeping eruption (Fig. 10). The wounds formed by scratching are frequently subjected to secondary bacterial invasion. The dog and cat hookworms, *A. braziliense* and *A. caninum*, are the principal causative agents of cutaneous larva migrans.

Ascaris lumbricoides

Ascaris lumbricoides is the largest intestinal nematode infecting humans. The parasite, commonly known as the "human roundworm", occurs endemically in many parts of the world, but is most prevalent in tropical and subtropical areas. The intensity of infection may reach staggering levels, with several hundred worms present in a single host. The adults live free and unattached in the small intestine for about one year, during which time the daily egg production per female may average 20,000. Female worms generally range from 20 to 40 cm in length. The smaller males measure 12-30 cm in length, and are readily distinguished from females by a ventrally curled tail. Unembryonated eggs pass in the feces and become infective in about three or four weeks. At moderate temperatures, the eggs will remain viable in moist soil for up to 5 years. Ingestion of contaminated salads and other foods resulting from indiscriminate defecation or the use of feces as a fertilizer is the common means of infection. Transmission of infection by windborne dust carrying eggs is also possible. When fully embryonated eggs are ingested, they hatch in the small intestine. The larvae burrow into the mucosa wall, enter the circulation and undergo an extensive migration to the liver, heart, and lungs. Upon reaching the lungs, the larvae break out of the pulmonary capillaries, enter the alveoli, ascend the bronchi, are swallowed, and eventually reach the small intestine where they develop into adults

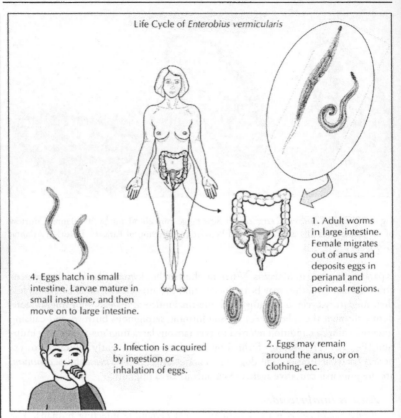

Life Cycle of *Enterobius vermicularis*

1. Adult worms in large intestine. Female migrates out of anus and deposits eggs in perianal and perineal regions.

4. Eggs hatch in small intestine. Larvae mature in small instestine, and then move on to large intestine.

3. Infection is acquired by ingestion or inhalation of eggs.

2. Eggs may remain around the anus, or on clothing, etc.

Figure 13. Life cycle of the pinworm, *Enterobius vermicularis*.

and mate. Adult worms begin laying eggs in the intestine about two or three weeks after infection. Transplacental migration of the parasite into a fetus is also (Fig. 11).

Symptoms of ascariasis are variable, often vague, mild or absent. Congestion, irregular respiration, spasms of coughing, edema, and a bloody sputum commonly characterize heavy pulmonary infections. Numerous parasites in the intestine may cause bowel obstruction, abdominal pain, nutritional and digestive disturbances, vomiting, restlessness, and disturbed sleep. Fatal peritonitis occurs rarely. The larvae may wander into such anomalous sites as the brain, spleen, liver, gallbladder, bile ducts, lymph nodes, peritoneal cavity, eustachian tubes, and the middle ear, causing inflammation, lesions, blockage of circulation or drainage, and death. The larvae may even exit from the body through the nasal passages. Live worms passed in the stool or vomitted are frequently the first manifestations of infection. Intestinal obstruction caused by the presence of numerous parasites may require surgical removal of the ascarids.

Pyrantel pamoate, piperazine hexahydrate or piperazine salts, levamisole, and thiabendazole are effective drugs for the treatment of ascariasis. Some side effects of

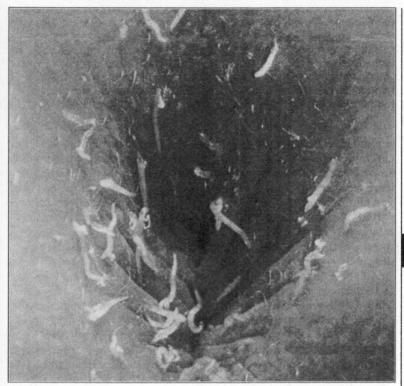

Figure 14. Pinworms in the perianal region. From Weber, M. Children's Hospital, Hannover Medical School, Hannover, Germany.

drug therapy of ascariasis include nausea, vomiting, diarrhea, and vertigo. When other intestinal worms are also present, *Ascaris* should be treated first to avoid the stimulation to migrate by the drugs.

Enterobius vermicularis

Enterobius vermicularis is the cosmopolitan pinworm, or seatworm, of humans. The parasite is the most common worm infection in the United States, with between 5 and 15% of the population harboring the parasite. Prevalence is highest in children of school age, and in individuals confined in institutions. The male worm is 2-5 mm in length, and females 8-13 mm in length. The posterior end of the male in strongly curved, while the female has a long, thin tapering tail. The adults characteristically congregate in the cecum and colon, but they do migrate within the lumen of the gastrointestinal tract from the stomach to the anus, feeding on the mucosa. Occasionally, adult worms in the stomach are regurgitated into the mouth. Gravid females migrate nocturnally out of the anus and deposit eggs on the perianal skin. The ovoid eggs are fully embryonated and infective within 6 hours after depo-

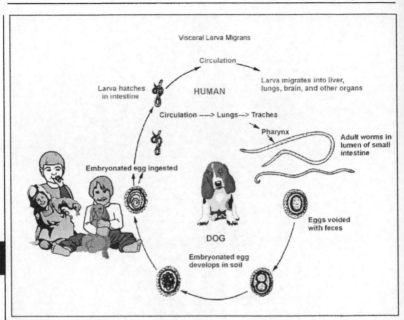

Figure 15. Life cycle of the intestinal roundworm of dogs (*Toxocara canis*) or cats (*T. cati*), agents that can cause visceral larva migrans in human.

sition (Fig. 12). Each female may deposit about 10,000 eggs, which can survive for a few weeks under conditions of high humidity and moderate temperature (20-25°C).

Infection results from direct transfer of infective eggs by hand from the anus to the mouth of the same or different host, or indirectly through contaminated food or articles. Viable eggs have been collected from clothing, bedding, towels, furniture, and house dust (Fig. 13). Infection by inhalation may occur in heavily contaminated environments. Domestic animals are free of pinworms, and thus cannot transmit the disease. When swallowed, the eggs hatch in the duodenum, and the larvae molt twice and mature in the ileocecal region in 2-6 weeks. Retroinfection occurs when the eggs deposited in the perianal region hatch and the larvae wander back into the intestine through the anus (Fig. 14).

The majority of pinworm infections are asymptomatic. Children with heavy worm infections often experience mild gastrointestinal irritation, nausea, vomiting, irritability, and restless sleep. Intense scratching of the perianal region may produce lesions and secondary bacterial infections. Inflammation of the intestinal mucosa occurs at times. Occasionally, pinworms migrate into the female genital tract and peritoneal cavity causing irritation and infection. Encapsulated worms have been found in the peritoneum, liver, and lungs, but these complications occur only rarely. Eggs inhaled or introduced into the nose by contaminated fingers may hatch, with the larvae surviving for a time in the nasal passages producing some irritation.

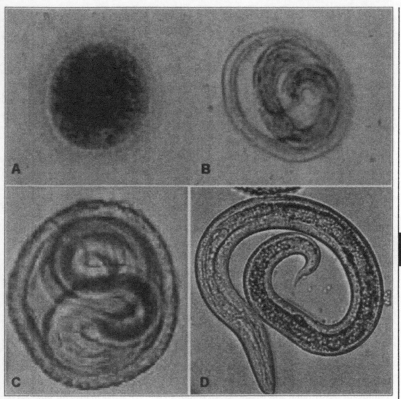

Figure 16. (A) Fertilized and unembryonated egg of *Toxocara canis* passed in dog feces. (B, C) Embryonated eggs of *T. canis*, each with a well developed larva. (D) Larva of *T. canis*. Courtesy of the Centers for Disease Control and Prevention, Division of Parasitic Diseases, Atlanta, Georgia.

Visceral Larva Migrans

Visceral larva migrans is a condition resulting from the accidental ingestion of the eggs of non-human nematoid larvae, the adults of which normally infect dogs and cats. The freed larvae are unable to complete their development in humans and wander erratically into various tissues and organs causing various pathologies. Frequently, the parasites become inactive and produce granulomas. Human visceral larva migrans frequently result from ingestion of the eggs of *Toxocara canis* or *T. cati*, two cosmopolitan intestinal parasites of domestic dogs and cats, respectively, and related carnivores. The disease, which is termed toxocariasis, is common in young children who always seem to have fingers going into their mouths (Fig. 15). Symptoms of infection typically include fever, eosinophilia, pulmonary infiltrates, and hepatomegaly. Granulomas have been found in the liver and retina. Death may result from extensive migrations of larvae, especially through the brain. In normal hosts, unembryonated eggs are passed in the feces and mature in the soil to the infective stage in about one week (Fig. 16). When ingested, the eggs hatch in the intestine, and the larvae burrow into the intestinal mucosa. The migratory route taken by the nematodes depends on the age and immunity of the host. In young dogs without previous exposure to the parasite, and thus with no established immunity against *T. canis*, the worms migrate through the portal system to the pulmonary circulation, up the trachea and back to the intestine. In mature dogs with some immunity from a previous infection, the larvae generally fail to complete the pulmonary migration and bypass the tracheal route leading to the small intestine. Instead, the larvae enter the systemic circulation and become distributed in various tissues and organs where they remain alive but inactive, encapsulated by a granulomatous host reaction. These dormant, encapsulated larvae appear to be activated by host hormones during pregnancy, and pass from the bitch into the fetal bloodstream to complete a pulmonary migration en route to the intestine. Parasite development occurs so rapidly that sexually mature worms may be found in young pups less than three weeks old. Except for the apparent absences of a lung migration and a prenatal infection route, the life cycle of *T. cati* is similar to that of *T. canis*. Rodents may serve as transport hosts when they ingest eggs of *T. canis* or *T. cati*. In rodents, the nematodes develop only to the second stage and become dormant in the tissues. It is estimated that 20% of the adult dogs and cats in the United States are infected with *Toxocara*. The infection rate may exceed 95% in pups and kittens.

Angiostrongyliasis

Two other nematode species causing human visceral larva migrans include *Angiostrongylus cantonensis* and *A. costaricensis*. The normal definitive hosts for these two species are rats. In humans, *A. cantonensis* causes eosinophilic meningoencephalitis. The disease is endemic in Hawaii, Tahiti, Japan, Taiwan, the Philippines, Australia, and Madagascar. Aquatic and terrestrial snails and slugs serve as intermediate hosts when they eat the feces of infected rats. Human infection is acquired by eating the infected intermediate or transport hosts, or raw vegetables or fruit containing viable parasite larvae expelled from these hosts. Symptoms of angiostrongyliasis may include severe headache and stiff neck (meningoencephalitis), marked eosinophilia in peripheral blood and cerebrospinal fluid, fever and coma. Although infection is usually benign, some fatalities have been reported. Temporary facial paralysis is occasionally observed. Human infections with *A. costaricensis* have been reported from Central America and Mexico. The terrestrial slug *Vaginulus plebeius* and various physid snails are intermediate hosts. Human infection results from eating contaminated vegetables. The worms characteristically cause abdominal pain and fever, thrombosis, inflammation and necrosis in the mesenteric arteries, and lesions in the walls of the small and large intestine. Occasionally, eosinophilia is found in infected individuals.

Anisakiasis

Anisakiasis is a disease of the gastrointestinal tract caused by ascarid larvae of the genus *Anisakis*. The adult nematodes are normally intestinal parasites of various marine fishes, birds, and mammals, such as whales, seals, dolphins, sea lions and porpoises. Many species of fish serve as transport hosts. The worms measure up to

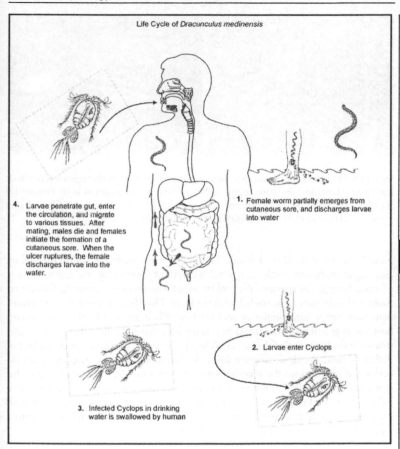

Life Cycle of *Dracunculus medinensis*

4. Larvae penetrate gut, enter the circulation, and migrate to various tissues. After mating, males die and females initiate the formation of a cutaneous sore. When the ulcer ruptures, the female discharges larvae into the water.

1. Female worm partially emerges from cutaneous sore, and discharges larvae into water

2. Larvae enter Cyclops

3. Infected Cyclops in drinking water is swallowed by human

Figure 17. Life cycle of the Guinea worm, *Dracunculus medinensis*.

3.5 cm in length. The larval stages are acquired by humans when raw or inadequately cooked infected hosts are eaten. The disease occurs predominately in Japan and adjacent areas, in various parts of Europe and Scandinavia, where raw, pickled, or smoked fish (herring, cod, salmon) are commonly eaten. The larvae do not mature in humans, but may cause acute nausea, vomiting, intestinal obstruction, abdominal symptoms mimicking those of appendicitis or ulcers, abscesses, gastric tumor-like growths, and peritonitis. Some infections have been reported to terminate fatally. There has been a marked increase in the incidence of anisakiasis in the United States due to the growing popularity of eating sushi.

Gongylonemiasis

Gongylonema pulchrum is a cosmopolitan parasite primarily of ruminants and swine, but the nematode has also been found in monkeys, hedgehogs, bears, and occasionally humans. Human infections have been reported in Europe, Russia, China,

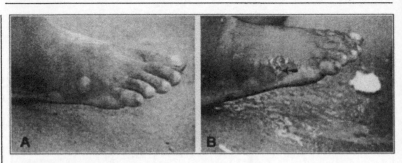

Figure 18. The female Guinea worm induces a painful blister (A) in the region where it will exit from the host. After the blister ruptures, the adult worm emerges (arrow in B). Frequently, the blister becomes secondarily infected. Courtesy of the Centers for Disease Control and Prevention, Division of Parasitic Diseases, Atlanta, Georgia.

New Zealand and the United States. The adult worms are found within the oral or esophageal epithelium. Males may reach a length of about 6 cm, and females about 15 cm in length. The eggs are fully embryonated when they pass from the females en route to the intestine to be voided with the feces. The life cycle involves an intermediate host, either dung beetles or cockroaches. When ingested by the intermediate host, the eggs hatch and the larvae penetrate the intestinal wall and become encapsulated in the hemocoel of the insect. When the intermediate host is ingested by the definitive host, the larvae migrate from the intestine up to the esophagus or oral cavity and burrow into the mucosa where they mature. In human infections the worms do not mature. The larvae may wander into the tongue, gums, hard and soft palate, lips and other areas of the oral cavity. Symptoms of gongylonemiasis include irritation, bleeding from the mouth, pharyngitis and stomatitis. The only effective treatment consists of surgical removal of the worms.

Gnathostomiasis

Gnathosoma spinigerum is the principal etiologic agent of human gnathostomiasis externa. The disease usually is acquired by humans through ingestion of raw or improperly processed or cooked freshwater fish infected with third-stage larvae of the nematode. The parasite is found in Japan, China, Thailand, Malaysia, Sumatra and the Philippines. Usually cats and dogs serve as definitive hosts. The adult worms are found embedded in tumor-like growths or nodules in the stomach wall. Male worms are 11-30 mm long, females 11-54 mm long. The eggs are unembryonated when passed in the feces. If deposited in water, the eggs mature in about one week and hatch. Two intermediate hosts are involved in the life cycle. The first intermediate host is a freshwater copepod that ingests the free-swimming parasite larva. The second intermediate host may be crustacea, freshwater fishes, amphibians, reptiles, birds, or mammals. When the infected copepod crustacean containing the second-stage larva is ingested by the vertebrate second intermediate host, the parasite penetrates the intestine of the new host, migrates to the connective tissues or muscles and molts to the infective third-stage larva. When the second intermediate host is

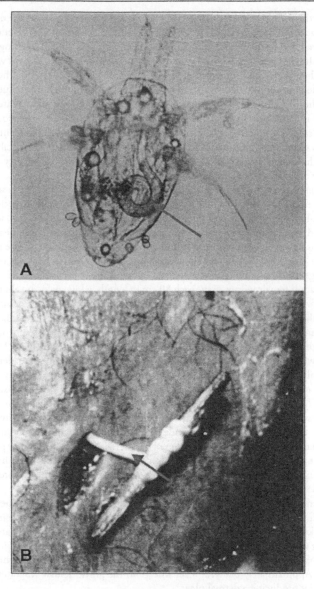

Figure 19. (A) Crustacean with larva of *Dracunculus* (arrow) in the body cavity. From Muller, R. 1971. *Dracunculus* and dracunculiasis. Advances in Parasitology (Ed. B. Danes). Volume 9, Academic Press, New York. (B) Guinea worm adult being removed from a leg by winding around a matchstick. From the Institute of Parasitology and Malariology, University of Teheran School of Medicine.

eaten by the definitive host, the parasite penetrates the wall of the stomach or intestine, invades the liver, muscles, and connective tissues. The worm later re-enters the gastric wall and becomes embedded in a nodule. Through a small opening in the nodule, the eggs pass into the lumen of the stomach and are then evacuated with the feces. When the second intermediate host is consumed by an unnatural host, the nematode penetrates the gut wall and migrates erratically in the host's tissues, usually in the subcutaneous layers, where they may produce endomatous areas and transient inflammatory lesions or abscesses which resemble those of cutaneous larva migrans caused by hookworms. Occasionally the worms will erupt from cutaneous abscesses spontaneously. The larvae may invade the brain producing cerebral lesions and death. Ocular damage and blindness also have been reported. Surgery is presently the only effective treatment for gnathostomiasis.

Dracunculus medinensis, or the guinea worm, is believed to be the "fiery serpent" reported in the Old Testament infecting Israelites during their travels into the Sinai Penninsula. The parasite is prevalent in India, Africa, and the Middle East, and it is estimated that about 48 million humans are presently infected. Guinea worm infections have also been reported from humans in the United States. The parasite in North America is probably a related species, *D. insignis*, which has been reported from dogs, raccoons, and other carnivores.

The adult female worms may reach 80 cm in length, while males are about 4 cm in length. The adults develop in the body cavity or visceral connective tissues (Fig. 17). Within a few months after copulation, the males die, become encapsulated and degenerate. The gravid female, which is ovoviviparous, migrates to the subcutaneous tissues, primarily of the extremities, where a dermal blister forms (Fig. 18). The blister, which develops where the head of the worm penetrates the dermis, is commonly found on the feet, ankles, calf, thigh, and knee joint, but it may also form on other areas of the body including the arms, trunk, buttocks, and scrotum. Eventually the blister ruptures, exposing the worm and a portion of the uterus, which has prolapsed through the body wall of the worm. When the blister is immersed in water, the worm and uterus protrude through the wound and numerous larvae are discharged. Emptied (spent) portions of the uterus disintegrate and new portions containing viable larvae move into the ulcer. After the uterus is emptied, the worm may be expelled, or it may withdraw and penetrate to deeper tissues before dying. Evacuation of the entire worm may take 3 weeks.

When ingested by a copepod the larvae penetrate the gut and enter the hemocoel of the crustacean where they molt twice and develop to the infective stage in about 2-3 weeks (Fig. 19). If the infected copepods are swallowed with contaminated drinking water, the larvae penetrate the intestine of the definitive host, and move via the lymphatics to the abdominal muscles and visceral connective tissues where they mature. Approximately one year after infection, the gravid female migrates to the extremities producing a dermal ulcer.

Although single infections of *Dracunculus* are most common, some individuals have been known to harbor as many as 50 worms. Migration of the gravid female to the extremities may initiate an allergic reaction, which causes erythema, urticarial rash, and intense pruritus. After several weeks, these symptoms usually subside or disappear. Female worms frequently fail to reach the surface and discharge larvae. Such non-emergent worms generally die and become absorbed or calcified, with no

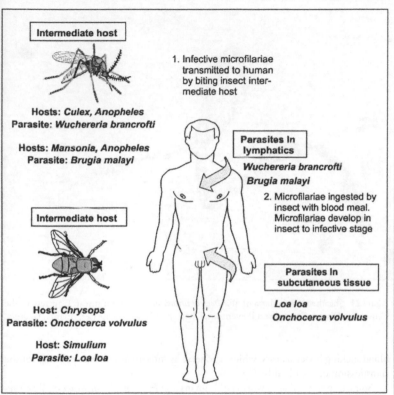

Figure 20. Generalized life cycles of filariids.

apparent adverse affect on the host. However, non-emergent worms may form pus-filled abscesses, which can lead to arthritis if located in or near a joint. The abscesses can be very large, containing numerous larvae and up to 0.5 liter of fluid. Rare complications result from adult worms in the urogenital system, thoracic cavity (causing pericardiasis), and central nervous system (causing paraplegia). Secondary bacterial infections may also cause serious complications.

The most common technique for removing a guinea worm is to slowly wind the parasite on a stick. Each day only a few centimeters of the worm are wound about the stick, with care exercised as not to rupture the worm and cause secondary infections (Fig. 19). Surgical removal of the worm after local anaesthetic is widely practiced in India and Pakistan.

Filarial Worms

Filariasis refers to infection with one of several species of filarial worm. Human filariids are slender, blood- and tissue-dwelling parasites that cause some of the most disfiguring and debilitating diseases known. Female worms do not lay eggs but instead, give birth to larvae termed microfilariae. The ingestion of microfilariae by

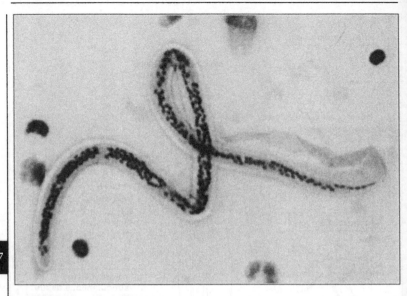

7

Figure 21. Sheathed microfilaria of *Wuchereria brancrofti* in a blood smear. Courtesy of the Centers for Disease Control and Prevention, Division of Parasitic Diseases, Atlanta, Georgia.

blood sucking insect vectors, which also serve as intermediate hosts, provides for the transmission of the filariids (Fig. 20).

Some important filarial worms of humans are: *Dracunculus medinensis, Wuchereria brancrofti, Brugia malayi, Loa loa, Onchocerca volvulus* and *Dirofilaria.*

Wuchereria brancrofti

Wuchereria brancrofti is the etiologic agent of Brancroftian filariasis or elephantiasis. The disease is the most widespread of the filariases of humans, occurring in Central and South America, Africa, Asia, West Indies, part of Europe and the Pacific islands. At one time the parasite was reported from humans living in the region of Charleston, South Carolina. Humans are the only known host for *W. brancrofti.* Female worms are 5 to 10 cm long, while males measure 2 to 4 cm in length. The adult worms live in tightly coiled masses in the lymphatic system where thousands of microfilariae are released by the females. The microfilariae eventually leave the lymphatic system and move into the blood via the thoracic duct. The microfilariae may be nocturnally periodic, being present in large numbers in the peripheral blood at night, and in low numbers during the day, or occur in greater concentration in the peripheral circulation in the daytime (diurnal periodicity). Variations in periodicity occur in different geographic strains of the parasite. Changes in host body temperature, activity, and in oxygen tension of the blood may be responsible for the filarial periodicity. Numerous species of mosquito belonging to the genera *Aedes, Culex, Monsonia* and *Anopheles* serve as intermediate hosts. The discovery by Manson in 1878 of the transmission of *W. brancrofti* by *Culex fatigens* in

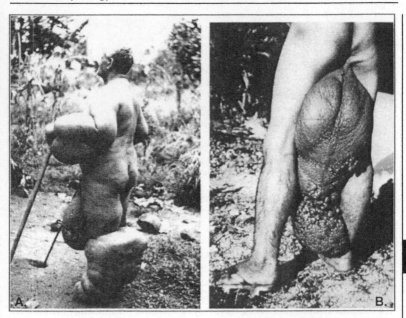

7

Figure 22. Extreme cases of elephantiasis of limbs and scrotum. (A) Photograph by Dr. John F. Kessel. From Markel, E. K., Vogue, M. and John, D. T. 1986. Medical Parasitology. W. B. Saunders Co., Philadelphia. (B) From Yamaguchi, T. (Ed.) 1981. A Color Atlas of Clinical Parasitology. (Wolfe Medical Publ. Ltd.) Lea & Febiger, Pennsylvania.

China was the first demonstration of an arthropod as a biological vector of a parasite (Fig. 20).

When the microfilariae are ingested by the mosquito along with a blood meal, the parasites penetrate the insect's gut, migrate to the thoracic muscles, and mature to the infective filariform stage. Infective larvae then migrate throughout the hemocoel and enter the proboscis of the mosquito, from which they escape when the insect again feeds. The larvae enter the skin of a new host through the puncture made by the feeding insect. The infective larvae move first to the peripheral lymphatics, then to the lymph nodes and larger lymphatics where they mature. Maturation of the worm may take up to one year from invasion of the skin until microfilariae appear in the blood (Fig. 21).

Clinical manifestations of Brancroft's filariasis result from the presence of adult worms in the lymphatic vessels. Light infections may go undetected, with the only physical signs of infection being slightly tender and swollen lymph nodes. Parturient females and microfilariae may cause intense lymphangitis accompanied by lymphadenitis, fever, chills, and toxemia. Lymphangitis commonly affects the arms and legs, and the epitrochlear and femoral lymph nodes are usually involved. Acute tissue reactions, consisting of hyperplasia of the vessel wall surrounding the parasite, and the accumulation of histocytes, eosinophils, and lymphocytes in the lumen of

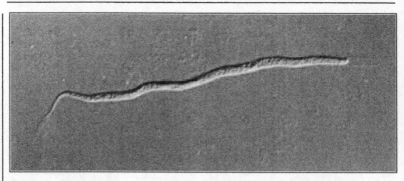

Figure 23. Microfilaria of *Brugia malayi* with a sheath extending beyond the anterior and posterior ends. Courtesy of the Centers for Disease Control and Prevention, Division of Parasitic Diseases, Atlanta, Georgia.

the vessels gradually obstruct the flow of lymph. Affected lymphatic vessels are distended, tender, and painful, and the overlying skin erythematous and hot. As affected areas become progressively infiltrated with fibrous connective tissue, a chronic lymphedema and thickening of the skin develops. Chyluria, or lymph in the urine, is a common symptom. In males, there may be extensive inflammatory involvement of the scrotum, with orchitis, epididymitis, and hydrocele (forcing of lymph into the tunica vaginalis of the testes or spermatic cord). In females, the vulva and breasts are sometimes affected. The development of elephantoid limbs and organs with dermal hypertrophy and verrucous growths represent uncommon and extreme manifestations of the disease, occurring in some individuals with repeated infections (Fig. 22). The swollen organs are composed primarily of fibrous connective tissue and fat. Many infected individuals never develop symptoms more severe than microfilaremia, localized edema, recurrent attacks of lymphangitis and fever, lymphadenopathy with splenomegaly, and transient pulmonary infiltrates and hypereosinophilia. Worms may be present for years in dilated or varicose lymphatic vessels before they die and are absorbed or become calcified. Occasionally, advanced cases of elephantiasis require surgery to correct lymphatic obstructions. Elephantoid tissues may also be treated by using elastic or pressure bandages, which force the lymph from the swollen areas. Surgical treatment of scrotal elephantiasis is commonly effective, whereas surgical treatment of elephantoid limbs is typically unsuccessful.

Brugia malayi

Brugia malayi is the etiologic agent of malayan filariasis. The parasite differs only slightly in morphology from *W. brancrofti,* and the life cycles of both species are nearly identical, with mosquitoes of the genera *Aedes, Mansonia* and *Culex* serving as intermediate hosts and vectors (Fig. 23). Females are 8 to 10 cm long. Males are about one-third the size of females. Malaysia is one of the major endemic areas, but the parasite also occurs in China, Korea, Japan, southeast Asia, Sri Lanka, the East Indies and the Philippines. Unlike *Wuchereria,* for which humans comprise the only known host, *B. malayi* also occurs in monkeys, dogs, and cats. The clinical and

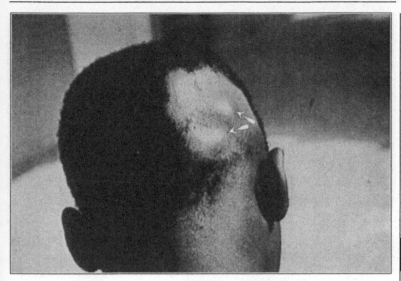

7

Figure 24. Cranial onchocercariasis. Adult *Onchocerca volvulus* larvae develop to adult worms in subcutaneous nodules (arrows). Courtesy of Professor Wallace Peters and the Carlo Denegri Foundation, Torino, Italy.

pathological features and treatment of malayan filariasis are similar to those described for the brancroftian variety. However, elephantiasis caused by *B. malayi* is typically not as severe as that caused by *Wuchereria*, and there is rarely any involvement of the genitalia.

Loa loa

Loiasis is a disease resulting from infection with *Loa loa*, the African eyeworm. The parasite, which is found only in Africa, is transmitted to humans by one of several species of mango or deer flies belonging to the genus *Chrysops*. Monkeys and other primates appear to be the only other definitive hosts. Female worms are 5 to 7 cm long, and males are approximately one-half the size of females. The sheathed microfilariae of *L. loa* undergo a developmental cycle in mango flies similar to that of *Wuchereria* in mosquitoes, with infective larvae passing from the fly's proboscis and entering the skin of the definitive host when the insect bites and draws blood. Following infection, the larvae develop into adult worms which migrate through the subcutaneous and deeper connective tissues of the definitive host. The microfilariae are periodic, appearing in large numbers in the peripheral circulation during the day, and in the lungs at night. Infections are generally asymptomatic, although occasionally migrating worms provoke localized inflammatory responses. Localized, transient subcutaneous nodules termed "calabar swellings" develop in areas where the worms remain stationary for brief periods of time. The swellings are believed to be allergic manifestations of the infection, and they generally disappear within a few days when the worms move to other areas. Clinical manifestations

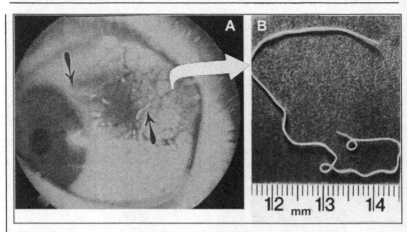

Figure 25. Ocular dirofilariasis caused by *Dirofilaria repens*. (A) A single worm coiled under the bulbar conjunctiva (small arrows) of the right eye. (B) Excised worm. Courtesy of Drs. Alberto Biglino and Angelo Casabianca, Asti General Hospital, Asti, Italy, and the Carlo Denegri Foundation, Torino, Italy.

result when the adults migrate into the facial area, conjunctival tissue, and cornea. When the worms are located in suitable superficial locations, they can be removed surgically, but this procedure is generally unwarranted because of effective chemotherapy.

Onchocerca volvulus

Onchocerca volvulus is the etiologic agent of onchocerciasis or river blindness, a disease found only in humans in central Africa, and northern South America. Female worms are 33-50 cm long by 0.2-0.4 mm wide. Males are 19-42 cm long by 0.1-0.2 mm wide. Adult worms are located in the dermis and subcutaneous tissues, where, as a result of host immune responses, they become enclosed in fibrous capsules that are usually visible as small nodules under the skin. The nodules (onchocercomas) develop on any part of the body, but they are frequently found on the trunk, hips, elbows, and scalp. Unsheathed microfilariae escape from the nodules and remain in the connective tissue of the skin. Rarely do the microfilariae enter the general circulation. Any one of several species of black fly belonging to the genus *Simulium* serves as the intermediate host and vector. The microfilariae are ingested by the biting flies when the insects feed on tissue fluids of the definitive host. The parasites penetrate the intestinal tract of the fly and move directly to the thoracic muscles, where they molt to the infective filariform stage. The parasites then pass to the mouthparts and enter a new host when the simuliid feeds again. Following introduction into the new host, the parasites migrate through the subcutaneous tissues and eventually become encapsulated.

Superficial nodules may be disfiguring, but generally are not painful (Fig. 24). Microfilariae migrating from the nodules throughout the dermis and connective tissues may cause sensitization reactions, severe pruritic dermatitis, ocular lesions, and blindness. In certain endemic regions of Africa, 30% of the population has impaired vision due to onchocercariasis. Thick, wrinkled, and hyperpigmented skin are also common symptoms. In extreme cases, infected skin around the pelvic region loses its elasticity, and a condition known as "hanging groin" or adenolymphocele occurs. The fold of skin, often containing lymph glands, may hang down to the knees. Infection with *O. volvulus* may also cause inguinal and femoral hernias. The parasite is also believed to cause pituitary damage and dwarfism in individuals living in certain parts of Uganda.

Dirofilaria immitis

Dirofilaria immitis, the heartworm of dogs, cats, foxes, and wolves, is widely distributed in the United States, and throughout most of the world. Adult worms are commonly found in the right ventricle and pulmonary arteries, but they also invade the subcutaneous tissues and eyes of the host (Fig. 25). Females are 2.5-3.1 cm long, and males measure 1.2-2.5 cm long. The microfilariae are nocturnally periodic. Mosquitoes belonging to the genera *Anopheles, Aedes* and *Culex,* and fleas, serve as intermediate hosts and vectors. Following ingestion by the insect, the larvae enter the cells of the malpighian tubules and develop to the infective stage. The larvae later migrate to the fat body and eventually reach the mouthparts, from which they escape and infect a new host when the insect again feeds. Infected animals may suffer from chronic endocarditis, pulmonary complications, enlargement of the liver, and inflammation of the kidney.

Dirofilaria immitis has been reported from humans several times, as has *D. tenuis* and *D. repens*, two filariids commonly found in the subcutaneous tissues of raccoons and dogs, respectively. The parasites have been found alive in the heart and lungs, and in solitary nodules or abscesses in various subcutaneous locations including the eyes, lips, arms, and breasts. The worms may reach sexual maturity in humans, but rarely do they produce microfilariae.

7

CHAPTER 8

Arthropods

Arthropods comprise approximately 78% of all known species of animals. They possess jointed appendages and a segmented body that is covered with a chitinized exoskeleton. There are two major groups of arthropods, chelicerates and mandibulates. The chelicerate arthropods include arachnids (spiders, scorpions, mites, and ticks), xiphosurids (king crabs), and pycnogonids (sea spiders). These organisms differ markedly from one another, but generally they possess chelicerae as the first pair of appendages, and they lack antennae. Mandibulate arthropods, which include insects, crustaceans, millipedes, and centipedes, typically possess antennae and mandibles, but lack chelicerae. Some of the characteristics distinguishing mandibulates and chelicerates are given in Table 1.

Arthropods are of medical and veterinary importance not only because they serve as intermediate hosts and vectors of disease, but also because they themselves are causal agents of disease. Some of the more important arthropod vectors and the diseases they transmit are given in Table 2. The remaining portions of this chapter consider arthropods as pathogenic agents.

Types of Injury Caused by Arthropods

The types of injury caused to humans by the adults and/or young stages of arthropods may be classified as follows:

1. Cutaneous or mucocutaneous injury. The development of blisters or lesions on the skin or mucous membranes may result from the discharge of body fluids (vesication), the contact with hairs, spines, or other cuticular projections (urtication), or the introduction of various secretions from biting or stinging arthropods (venenation).

2. Systemic injury. Various pathologies result from the introduction of secretions into the body by venenating arthropods, including parasitemia, fever, hemorrhage, tachycardia, and respiratory difficulties.

3. Tissue injury. Mechanical damage to visceral organs and tissues frequently results from the penetration of the body by the adult or young stages of arthropods. Blood-sucking arthropods have been classified into two groups according to the method used to obtain blood. Solenophagic feeders insert their mouthparts directly into the lumen of a blood capillary and thus feed almost exclusively on blood. Adult mosquitoes are solenophagic feeders. Telmophagic feeders insert their mouthparts indiscriminately into the skin. The enzymatic secretions initially cause lysis of the tissues, and the animal feeds first on tissue fluids until the walls of the dermal blood vessels are broken down and a blood pool is produced from which they feed. Ticks are good examples of telmophagic feeders.

Parasites of Medical Importance, by Anthony J. Nappi and Emily Vass.
©2002 Landes Bioscience.

Table 1. Some general characteristics of the major groups of arthropods

	Mandibulates				Chelicerates
	Crustaceans	Insects	Centipedes	Millipedes	Arachnids
Body divisions	Usually cephalothorax and abdomen	Head, thorax, abdomen	Head and trunk	Head, thorax, abdomen	Cephalothorax and abdomen
Mouth parts	Mandibles, maxillae (2 pairs)	Mandibles, maxillae (1 pair)	Mandibles, maxillae (2 pairs)	Mandibles, maxillae (1 pair)	Chelicerae, pedipalpi
Legs	1 pair per segment, or less	3 pairs on thorax	1 pair per segment	2 (or 1) pairs per segment	4 pairs on cephalothorax
Antennae	2 pairs	1 pair	1 pair	1 pair	None
Respiration	Gills or body surface	Tracheae	Tracheae	Tracheae	Book lungs, or tracheae
Genital openings	2, hind part of thorax	1, end of abdomen	1, end of abdomen	1, third segment near head	1, second segment of abdomen
Development	Usually with larval stages	Usually with larval stages	Direct	Direct	Direct, except mites and ticks

Arthropods as Vectors of Disease

Arthropods are of medical and veterinary importance not only because they serve as intermediate hosts and vectors of disease, but also because they themselves are causal agents of disease. A list of some of the more important arthropods and the diseases they transmit is given in Table 1. The remaining portions of the chapter consider arthropods as pathogenic agents.

Chelicerates (Arachnids)

Scorpions

Scorpions are the largest of the arachnids, measuring up to 20 cm in length. They are cosmopolitan in distribution through tropical and subtropical regions. Scorpions are largely nocturnal carnivores feeding on insects, centipedes, spiders, and small mammals. The abdomen terminates in a sting that can penetrate the integument of the prey, into which venom is injected (Fig. 1). The venom of most scorpions is harmless to humans. The wound produced by the sting is painful, and is usually accompanied by local swelling and discoloration. A radiating, burning sensation is generally experienced at the time of venenation. A few species (*Centruroides, Androctonus, Parabuthus, Buthacus, Buthotus, Scorpio* and *Tamulus*) are capable of inflicting systemic reactions, which may cause death. Young children are most liable to venenation and experience a high mortality. The venom affects the nervous system and also causes pulmonary disorders. Symptoms may include fever, vomiting, frothy salivation, extreme thirst, muscle spasms, convulsions, impaired respiration, tachycardia, paralysis and death. Early therapeutic measures include placing an ice pack over the wound, the application of a tourniquet proximal to the site of venenation, and the removal, by suction, of as much venom as possible. If available, anti-venom should be administered in severe cases involving systemic infections.

Spiders

Spiders rank among the most feared of animals, an emotion that is unjustified, since the vast majority of species are completely harmless to humans. The body of a spider is comprised of two distinct regions, an anterior cephalothorax and a posterior abdomen (Fig. 2). The cephalothorax bears four pairs of walking legs, a pair of pedipalps, and a pair of chelicerae. One to four pairs of silk-spinning organs or spinnerets occur near the terminal portion of the abdomen. Most spiders have eight eyes arranged in two rows of four. The chelicerae are comprised of two segments, a basal portion and a distal fang, which is used to puncture prey and introduce venom. Spiders feed on insects, other arthropods, and in the case of some tropical spiders, small vertebrates. Following the initial puncturing by the chelicerae, the prey is subjected to extra-oral digestion by enzymes secreted from the mid-gut, and from glands found in the pedipalpal coxae. The liquid tissues are then sucked from the prey by means of the pharynx.

Few spiders are medically important. Many attacks on humans occur only accidentally when an exposed part of the body comes in contact with the spider. Depending upon the species of spider, two types of pathophysiological disorders may be produced, necrotizing lesions or systemic injury. The most important spiders causing necrotic arachnidism belong to the genus *Loxosceles*, and are commonly known as "brown" or "violin" spiders, since some species have a fiddle-shaped marking on the dorsal surface of the abdomen. *Loxosceles laeta* is widely distributed in southern South America. *Loxosceles reclusus*, the brown recluse spider, is the chief agent of loxoscelism in the mid-western United States. The bite of the spider is painful, and swelling and tissue necrosis usually develop. In mild loxoscelism, the cutaneous lesion heals leaving a disfiguring scar several centimeters in diameter. Manifestations of systemic injury may include fever, hemorrhage, erosion of mucous membranes, hematuria, and cardiac failure. Corticosteroids may alleviate the pain and prevent the development of necrosis and systemic damage.

Many of the spiders producing systemic injury belong to the genus *Latrodectus*. The "black widow", *L. mactans*, is distributed in Canada, the United States, and parts of South America. Other species of *Latrodectus* occur in these areas, as well as in the Mediterranean area, Africa, New Zealand, parts of the Caribbean, Asia and Europe (Fig. 3). The venom acts as a peripheral neurotoxin. Immediate sharp pain occurs at the site of the bite, which later becomes reddened. Symptoms include pain in the abdomen and chest, weakness, motor disturbances, spastic contractions, delirium, and convulsions. Death from latrodectism is rare, but may occur in young children or the elderly, and frequently results from respiratory and/or circulatory failure. Analgesics and sedation generally suffice as treatment for most cases. In se-

Table 2 Some human diseases transmitted by arthropods

Disease	Pathogen	Vector	Vertebrate Reservoir
Bacterial, Rickettsial, and Related Diseases			
Bubonic plague	*Pasteurella pestis*	Fleas, especially the rat flea	Rodents
Tularemia	*P. tularensis*	Deer flies, ticks, fleas, body lice	Rabbits
Anthrax	*Bacillus anthracis*	Horse flies	Mammals
Epidemic typhus	*Rickettsia prowazekii*	Body louse, rat flea, rar mite	Rodents
Endemic or murine typhus	*R. prowazekii, R. mooseri*	Fleas, lice, mites, rodent ticks	Rodents
Scrub typhus or tsutsugamushi	*R. tsutsugamushi*	Harvest mites (chiggers)	Rodents
Spotted fever	*R. rickettsii*	Ticks	Rodents
Q-fever	*Rickettsia* spp.	Ticks	Cattle, sheep, goats
Bartonellosis or oroyo fever	*Bartonella bacilliformis*	Sand flies	?
Protozoan Diseases			
African Trypanosomiasis	*Trypanosoma gambiense, T. rhodesiense*	Tsetse flies	Herbivores
South American Trypanosomiasis	*Trypanosoma cruzi*	Assassin bugs	Rodents, carnivores
Epizootic (Leishmaniasis)	*Leishmania braziliensis*	Sand flies	Dogs, cats
Oriental sore (Leishmaniasis)	*L. tropica*	Sand flies	Dogs, cats
Kala-azar (Leishmaniasis)	*L. donovani*	Sand flies	Dogs, cats
Malaria	*Plasmodium vivax, P. malariae, P. falciparum, P. ovale*	Mosquitoes	None

continued on next page

vere cases, the intravenous injection of specific antivenom, if available, should be administered.

Mites and Ticks (Acarina)

The order acarina is comprised of mites and ticks. Ticks differ from mites in their comparatively large size and their exclusive habit of feeding on vertebrate blood during each of their developmental stages. The acarine body is comprised of a head region (capitulum or gnathosoma), which bears the mouth parts, and a posterior region (idiosoma), which is subdivided into a leg-bearing part and an abdominal part. The first developmental stage of is a six-legged (hexapod) larva that emerges from the egg. With one or more intervening molts, the larva becomes an eight-legged nymph, which after one or more molts, becomes an adult.

Table 2. Cont.

Disease	Pathogen	Vector	Vertebrate Reservoir
Viral Diseases			
Yellow fever		Mosquitoes (*Aedes aegypti*)	Monkeys
Dengue fever		Mosquitoes (*Aedes aegypti*)	?
Encephalitis		Mosquitoes (spp. *Culex, Aedes, Mansonnia*)	Horses, birds
Pappataci fever		Sand fly	Humans
Colorado tick fever		Ticks	Rodents
Spirochaetal Diseases			
Relapsing fever	*Borrelia recurrentis*	Ticks	Rodents
Relapsing fever	*B. duttonii*	Body louse	None
Fluke Diseases			
Lung fluke	*Paragonimus westermani*	Fresh-water crabs, crayfish	Carnivores
Tapeworm Diseases			
Dipylidiasis	*Dipylidium caninum*	Dog flea	Dogs, cats
Hymenolepiasis	*Hymenolepis diminuta*	Rat flea, grain beetles, earwigs	Rats, mice
Diphyllobothriasis	*Diphyllobothrium latum*	Water fleas	Dogs, bears
Roundworm Diseases			
Brancroftian filariasis	*Wuchereria brancrofti*	Mosquitoes (*Aedes, Culex, Anopheles, Mansonia*)	?
Malayan filariasis	*Brugia malayi*	Black flies	Felines, monkeys
Loaiasis	*Loa loa*	Deer flies	Monkeys
Onchocerciasis	*Onchocerca volvulus*	Black flies	None
Guinea worm	*Dracunculus medinensis*	Water flea	Carnivores

Mites

Mites usually measure less than 1 mm in length. The chelicerae and pedipalps are located on the capitulum. A median hypostome is situated below the mouth. In scavenging and predatory mites the chelicerae are chelate and used for prehension. In parasitic forms, the chelicerae are stylet-like. The chelicerae open an incision, and the hypostome penetrates into the tissues. In some mites, external digestion results from secretions from the paired salivary glands. The liquid food is drawn into the

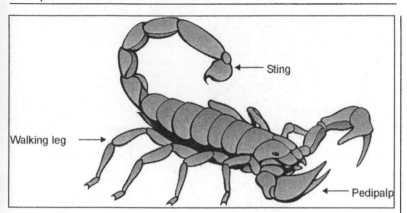

Figure 1. Scorpion.

pre-oral food canal by the suctorial action of the pharynx and then passes to the esophagous and gut. A few mites, such as mange and follicular mites, burrow into the tissues of the host.

Red bugs or "chiggers" are larval mites of the family Trombiculidae. Only the larval stage is parasitic, sucking tissue juices from various vertebrates, including humans. Upon hatching from eggs laid on the ground or vegetation, the hexapod larvae attach to the skin of vertebrates, inject a salivary secretion that lyse the integument, and suck tissue fluids until engorged. After feeding, which may last from a few days to several weeks, the larvae drop to the ground and molt to become free-living nymphs and finally adults, all of which feed on the eggs of insects. Trombiculids cause a type of dermatitis, which may be characterized by pruritus and bleeding. Areas frequently attacked are the ankles, legs, groin, external genitalia, waistline, axillae and breasts. The common species of trombiculids in the United States belong to the genus *Trombicula*. In parts of Asia, larvae of *T. akamushi* and related species serve as biological vectors of *Rickettsiae tsutsugamushi*, the etiologic agent of scrub typhus or tsutsugamushi fever. The disease is transmitted by the mites from rodent reservoir hosts to humans.

Several species of mites are known as mange mites. They infest the skin of mammals causing severe irritation, cutaneous lesions, and mange by burrowing into the skin. Psoroptic mange or "scab" is caused by mites belonging to the genera *Psoroptes* and *Chorioptes*. These mites feed on the surface of the skin, and remove portions of the epidermis as they spread upon the raw dermal layer under the scabs which form. Sarcoptic mange and "itch" is caused by burrowing mites. The "itch mite", *Sarcoptes scabiei*, forms cutaneous tunnels a few millimeters to several centimeters in length, and feeds on host tissues (Fig. 4). The itch mites of cattle, hogs, sheep, dogs and humans are believed to be varieties of this species. The adults, which range in length from 0.2 to 0.45 mm, are found on the surface of the skin at night, and can easily be transferred by personal contact and by clothing to new hosts. Gravid females deposit eggs and feces as they excavate tunnels in the epidermis. Three to five days after oviposition, the eggs hatch and the hexpod larvae emerge and either burrow off the

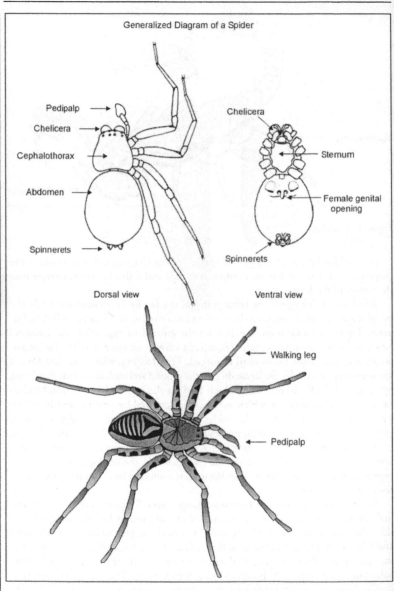

Generalized Diagram of a Spider

Figure 2. Morphology of a typical spider.

main channel or exit and form new tunnels. The larvae molt to become nymphs, and later adults. The life cycle is completed in one to two weeks.

The initial infection is generally mild and asymptomatic, but after a few weeks the disease spreads and is accompanied by intense itching. Scratching causes lesions

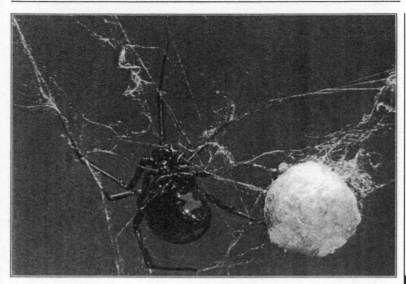

Figure 3. Black widow spider, *Latrodectus mactans*. From Knopf, A. A. 1997. National Audubon Society: Insects and spiders. Chanticleer Press, Inc., New York.

to form in the dermis, and secondary bacterial infections to develop. The lesions are most commonly found between the fingers and toes, on the back of the hands, in the groin, axillary regions, and external genitalia.

Demodex folliculorum, or follicle mite, is a small (0.1 to 0.4 mm), elongated, worm-like parasite that burrows into the hair follicles and sebaceous glands of humans (Fig. 5). The burrows are shallow, and the mites themselves generally produce only a mild pruritus, acne, or local keratitis. Demodectic mange in dogs results from bacterial complications following invasion of the skin by mites.

Various species of mites live on poultry and rodents. Those of medical and veterinary interest belong to the genera *Dermanyssus, Liponyssoides, Allodermanyssus* and *Ornithonyssus*. The common red chicken mite, *D. gallinae*, is found on various types of foul, and occasionally attacks humans. The mite is a nocturnal feeder, sucking blood and tissue fluids of the host. The introduction of toxic salivary secretions into the skin during feeding produces a severe dermatitis and papular eczema. The mite harbors viruses that cause St. Louis encephalitis and western equine encephalitis. The mouse mite, *L. sanguineus*, also produces a severe dermatitis in humans. In addition, the mite is the vector of *Rickettsiae akari*, the etiologic agent of human rickettsial pox. Mites transmit the disease from the reservoir host, the house mouse (*Mus musculus*), to humans. The rat mite, *O. bacoti*, and the tropical fowl mite, *O. bursa*, also produce an annoying dermatitis in humans. In addition, *O. bursa* transmits western equine encephalitis, and *O. bacoti* vectors *Rickettsia typhi* (the cause of endemic typhus), *Rickettsia akari* (the etiologic agent of rickettsial pox), and the virus that causes Q fever.

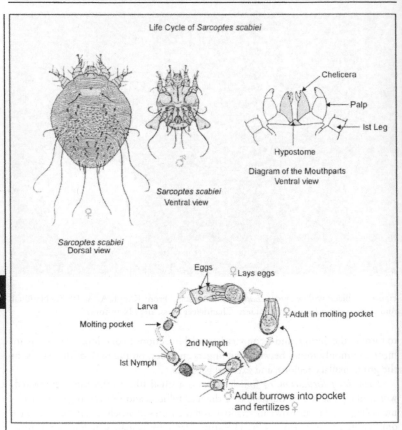

Figure 4. Life cycle of the itch mite, *Sarcoptes scabiei.*

Various species of itch mites are agricultural pests that inhabit warehouses and grain mills, or are predaceous on insects which infest grain crops. Dermatitis and allergic reactions variously termed "grocer's itch", "copra itch" or "miller's itch", result when humans come in contact with mite-infested plant products such as cereals, grains, cheese, dried fruits, and sugar. The mites generally crawl under the epidermis and produce temporary pruritism, but some have been found in the lungs, ears, intestinal and urinary passages.

Ticks

There are two distinct types of ticks, hard or ixodid ticks, and the soft or argasid ticks. Ixodid ticks have a dorsal shield or scutum covering virtually the entire back in the males, and only a small portion anteriorly in the females. Argasid ticks lack plates or shields, and have a leathery cutical. In ixodid ticks, the mouth is anterior in position, and the adults feed only once. In argasid ticks, the mouth is ventral, and the adults feed continually. The eggs are deposited on the ground, and the young hexapod larvae, termed "seed ticks", climb up on vegetation and attach to their

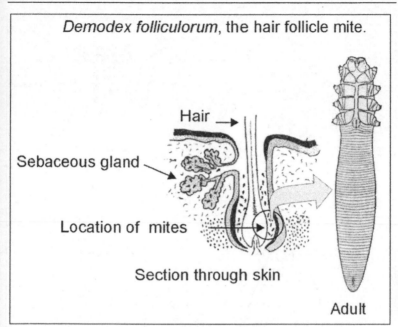

Demodex folliculorum, the hair follicle mite.

Figure 5. The hair follicle mite, *Demodex folliculorum*.

hosts on contact with them. The young tick makes an incision into the skin with the cheliceral digits, and the chelicerase and hypostome are then inserted into the wound (Fig. 6). The barbed hypostome does not aid in skin penetration, but prevents the animal from being dislodged from the skin. In addition to producing a histolytic secretion, which liquifies the host tissues, some species produce a salivary secretion, which hardens to a latex-like consistency around the embedded mouth parts, preventing the tick from being removed during the feeding period. The liquid food is sucked into the gut by the action of the muscular pharynx. After feeding for a few days, the larvae drop to the ground and molt to become eight-legged nymphs. The nymphs, in turn, will attach to their host and again feed. After feeding, the nymphs will return to the soil and molt to the adult stage. The adult ticks repeat the process, attaching themselves to their hosts and feeding.

Ticks are of considerable medical importance primarily because of their role as vectors of diseases caused by bacteria, viruses, and rickettsia. The argasid ticks of medical importance belong to the genera *Ornithodorous, Otobius* and *Argas.* Unlike the hard ticks, the argasids are nocturnal feeders, which hide in various cracks and crevices during the day. The medically important ixodid ticks belong to the genera *Dermacentor* and *Amblyomma.* In addition to being vectors of diseases, saliva introduced by some species into the wound during feeding may cause a type of sensitization reaction, which is systemic. Much more serious is tick paralysis, which at times is fatal. The paralysis appears to be an occasional host reaction to a toxin presumably elaborated from the eggs of feeding female ticks. Apparently, it occurs

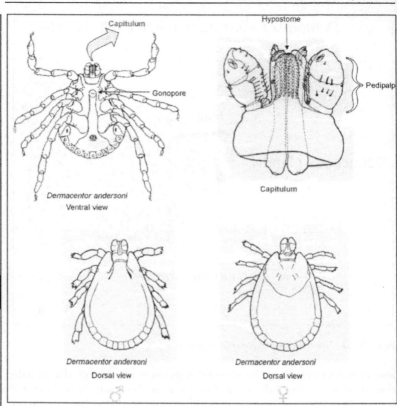

Figure 6. The ixodid tick, *Dermacentor andersoni*.

most frequently when ticks feed for extended periods of time at the base of the skull or back of the neck and along the spinal column. The ixodid ticks, which cause paralysis, are *D. andersoni* in western North America, and *D. variabilis, A. americanum* and *A. maculatum* in eastern and southern United States. The paralysis is characterized by ascending motor paralysis, which originates in the legs and moves up to include the chest and neck, and impairment of respiration, speech and swallowing. There is no specific treatment for tick paralysis. Paralysis terminates and recovery follows if the tick is removed before the heart and respiration are affected. Secondary infections with lesions and systemic poisoning may result if the capitulum is broken off when improper attempts are made to remove the embedded tick. The application of various oils or creams to the body of the tick generally initiates its withdrawal from the skin of the host.

Insects

Insects comprise about 70% of all known species of animals. Both economically and medically, insects represent the most important group of animals. The orders of insects containing species of medical importance include Coleoptera (beetles),

Orthroptera (cockroaches), Hymenoptera (ants, bees), Anoplura (sucking lice), Hemiptera (true bugs), Siphonaptera (fleas) and Diptera (flies, mosquitoes, midges).

Coleoptera (Beetles)

Blister beetles produce a vesicating fluid, which is present in their body tissues. When the beetle is accidentally crushed on the skin, the substance causes irritation and blistering at the site of contact. A common vesicating beetle is *Lytta vesicatoria* or "Spanish fly". The cuticular hairs or small body fragments of some dermestid beetles may act as allergens producing asthma-like reactions when inhaled. The accidental ingestion of infected beetles, which serve as intermediate hosts of helminths can produce infections in humans. Examples include the tapeworm parasite of rodents, *Hymenolepis diminuta*, and the intestinal roundworm of pigs and ruminants, *Gongylonema pulchrum*.

Orthroptera (Cockroaches)

Cockroaches are extremely common wherever food or garbage is found. In addition to being effective transmitters of enteric pathogens, such as *Entamoeba histolytica*, they also serve as intermediate hosts for the cestode *Hymenolepis diminuta*, and the nematode *Gongylonema pulchrum*.

Hymenoptera (Ants, Bees)

Wasps and certain other Hymenoptera affect humans adversely chiefly by injecting venom into the skin. In some species the modified ovipositor of the female serves as a stinging apparatus (Fig. 7). The sting may cause only local irritation, but some species can inflict serious injury. The host reaction to the venom may be immediate or it may be delayed for hours. The fire ant, *Solenopsis* sp., which is found in the southern United States, produces a very painful, burning sensation following its sting. Some individuals become sensitized to components of certain bee and wasp venoms, and subsequent stings may cause fatal anaphylactic shock.

Anoplura (Sucking Lice)

Sucking lice are small, dorsoventrally flattened ectoparasites of mammals (Fig. 8). The mouthparts of these insects are adapted for piercing and sucking, and their legs are modified for clasping hairs or fibers. Lice are not only host-species specific, but also site specific. All developmental stages live permanently on the clothing or body surfaces of their hosts to which the eggs are cemented. Louse infestation (pediculosis) is most common among individuals living in crowded environments, such as military camps, prisons, and mental institutions, and where facilities for personal hygiene are poor or lacking. The human pubic louse, *Phthirus pubis*, attaches its eggs to the hairs of the pubic region, chest, eyebrows and eyelashes. The body louse, *Pediculus humanus*, occurs as two habitat-specific strains. One strain, designated *P. humanus capitis* (head louse), lives only on the head, attaching its eggs ("nits") to the hair of the head or neck (see Fig. 8). The second strain, *P. humanus corporis* (body louse or "cooties"), cements its eggs to clothing fibers. The two strains are not completely genetically isolated, and can interbreed easily. Saliva introduced by lice into the skin causes pruritus, and scratching of the lesions results in an eczematous dermatitis. *Pediculus humanus corporis* and *P. humanus capitis* are vectors of the organisms causing epidemic typhus, trench fever, and relapsing fever. Transmission

Figure 7. The insect order hymenoptera includes wasps, bees, and ants. In some species, the ovipositor is modified as a sting. From Knopf, A. A. 1997. National Audubon Society: Insects and spiders. Chanticleer Press, Inc., New York.

of the pathogens occurs when infected lice are crushed on the skin, or by venenation when the insects feed.

Hemiptera (True Bugs)

The order Hemiptera is comprised of insects known as true bugs. The most conspicuous external feature of these avid blood suckers is the proboscis, which is used to pierce tissues and suck fluids (Figs. 9 and 10). Two medically important families are the Cimicidae (bedbugs), and the Reduviidae (triatomids or assassin bugs). Bedbugs are dorsoventrally flattened, bloodsucking ectoparasites of birds and mammals. The two most common species attacking humans are *Cimex lectularius* of temperate climates, and *C. hemipterus* of the tropics. The former species occurs also on chickens, rabbits and bats. The adults and immature stages of bedbugs may be found in cracks and crevices, or under carpeting during the day, and emerge at night to feed. Inflamed, cutaneous lesions may be produced at the puncture sites, accompanied by systemic sensitization. The reduviids which attack humans and other vertebrates belong to the genera *Rhodnius*, *Triatoma* and *Panstrongylus*. Some assassin bugs commonly bite on the lips and near the eyes producing painful wounds and localized bleeding. Many reduviids also vector *Trypanosoma cruzi*, the etiologic agent of Chagas' disease.

Siphonaptera (fleas)

Fleas are ectoparasites with laterally compressed bodies. They are important as pests of humans and domestic animals and as vectors of disease. Fleas generally are not very host-specific, and unlike lice and acarines, fleas have a non-parasitic larval stage. Female fleas deposit eggs on the ground near sleeping areas of their hosts. The larvae, which have chewing mouthparts, feed on organic waste and fragments of hair and epidermis of the host. The adults have mouthparts adapted for piercing and sucking, and long, powerful legs providing them with excellent jumping ability (Fig. 11). The fleas of medical importance include the human flea (*Pulex irritans*), the dog flea (*Ctenocephalides canis*), the cat flea (*C. felis*), and the fleas of rats, mice, and other rodents (including *Xenopsylla cheopis*, *Nosopsyllus fasciatus* and *Leptopsylla*

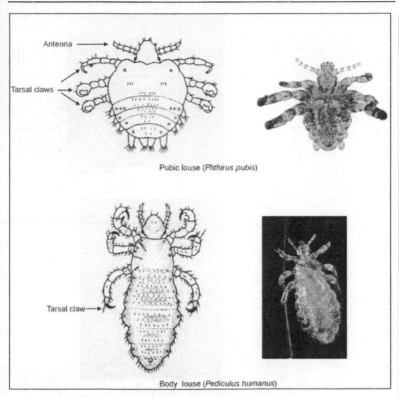

Pubic louse (*Phthirus pubis*)

Tarsal claw →

Body louse (*Pediculus humanus*)

Figure 8. Sucking lice *Phthirus pubis* and *Pediculus humanus.*

segnis). In some humans, fleas cause a local sensitization reaction at the site of infestation. The lesions may become secondarily infected as a result of scratching. Fleas from dogs and cats frequently produce a dermatitis when they attack humans. The dog flea also serves as an intermediate host of the dog tapeworm, *Dipylidium caninum*, which also can be transmitted to humans. The cysticercoid of *Hymenolepis diminuta* develops in larvae of several species of fleas. Rodent fleas transmit two serious diseases, bubonic plague and endemic typhus. The sticktight flea, *Echidnophaga gallinacea*, lives partly embedded in the skin of poultry, usually on the head near the eyes and bill. Occasionally the flea invades the ears of dogs and cats, and may also attack children. In addition, the flea can transmit plague among birds and rodents.

The sand flea or chigoe, *Tunga penetrans*, lives in sandy soil in tropical and subtropical America and parts of Africa. Unlike other fleas, which puncture the skin of humans and suck blood, *T. penetrans* invades the skin, usually burrowing under the toenails. The females engorge themselves by feeding on tissue fluids, and their abdomens become enormously distended. The fleas attach to a wide range of hosts, but are most commonly found on hogs and dogs. Humans walking barefoot on contaminated soil easily become infected. The parasite produces a painful sore which frequently becomes inflamed and secondarily infected and may result in lameness to

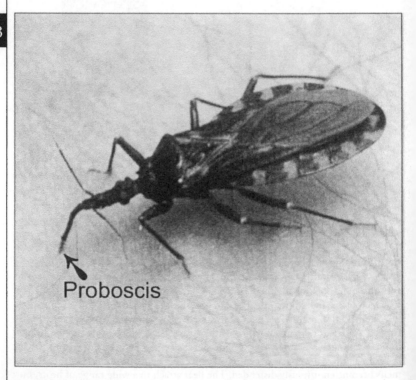

Cimex lectularius (bedbug)

Proboscis — Dorsal aspect

Proboscis — Antenna — Ventral aspect

Bedbug *on skin*

Figure 9. The bedbug, *Cimex lectularius.*

8

Proboscis

Figure 10. A reduviid bug.

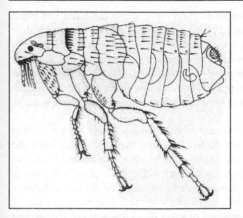

Figure 11. A flea.

the host. Death from gas gangrene and tetanus have been reported as complications of chigoe infestation.

Diptera (Flies, Mosquitoes, Midges)

Insects belonging to the Order Diptera are the most medically important arthropods. They are important ectoparasitic bloodsuckers that serve as biological and mechanical vectors of numerous diseases, and their larvae (maggots) are causal agents of myiasis. The adults of most species are characterized by having only one pair of functional, membranous wings, the forewings. The posterior wings are reduced to knoblike or club-shaped structures termed halteres which, in some species, are known to function as organs of balance.

The Family Psychodidae is comprised of sand flies (*Phlebotomus*), which are predominately nocturnal feeders. Only the females bite and suck blood. The insects produce an irritating pruritus at the site of attack. Allergic reactions develop in some hosts. Sand flies are biological vectors of the organisms that cause Leishmaniasis and bartonellosis (verrura peruana or Oroya fever).

The Family Ceratopogonidae is comprised of various small insects known as biting midges ("punkies", "gnats" and "no-see-ums"). Only the females puncture the skin and feed on blood. The majority of medically important species belong to the genus *Culicoides*. They usually attack humans at dusk, inflicting painful bites which may be followed by severe pruritus. Some species transmit to humans and other vertebrates the filariids *Dipetalonema perstans* and *D. streptocerca*.

Mosquitoes belong to the Family Culicidae. They are not only serious pests, but also important vectors of disease. *Anopheles, Aedes, Culex, Mansonia* and *Psorophora* are among the more important genera affecting humans. Some of the major diseases transmitted by mosquitoes include malaria, filariasis (caused by the nematodes *Wuchereria brancrofti* and *Brugia malayi*), dog heartworm (caused by *Dirofilaria immitis*), dengue fever, equine encephalitis, west Nile fever, and yellow fever.

The Family Simulidae is comprised of black flies, or buffalo gnats. The adults have an elongated thorax (hump-back) and rasping mouthparts. They are diurnal feeders, which attack various birds and mammals. They feed on the abraded tissues, blood, and tissue fluids of their hosts. Only the female black flies suck blood. The adults are generally found close to streams and lakes, where the larval and pupal stages develop attached to rocks or to plants underwater. The females attach their eggs to objects at, or just below, the water surface. Occasionally, the adults occur in tremendous swarms causing considerable annoyance to humans and domestic animals. The bites may produce intensely pruritic lesions. Hypersensitization may occur in persons that are attacked repeatedly by black flies. Some species of *Simulium* are biological vectors of the medically important filarial worm *Onchocerca volvulus*, which causes river blindness.

Members of the Family Tabanidae are robust flies commonly known as horse flies, deer flies, or mango flies. Only the females pierce the skin of their hosts and suck blood. They are mostly diurnal species, which inflict very painful bites. Two bacterial diseases, tularemia (*Pasteurella tularensis*) and anthrax (*Bacillus anthracis*), are transmitted to humans and other animals by members of the genus *Tabanus*. *Chrysops*, which also carries tularemia, is the vector of the filarial worm *Loa loa*.

The family Chloropidae is comprised of small flies known as eye flies or fruit flies. The flies cluster about the eyes, mucous membranes, and open sores of humans and domestic animals. The adult flies are frequently encountered in sandy or moist areas, and near irrigated and cultivated farm lands. Although incapable of piercing the skin, they do produce small lesions with their spinose labella. These flies are believed to transmit pinkeye (acute contagious conjunctivitis) and yaws, a painful and disfiguring disease caused by the spirochaete *Treponema perenue*. The larval stages of eye flies breed in moist soil and in feces. The eye gnat, *Hippelates pusio*, occurs in the southern United States. A related species, *H. pallipes*, is common in the West Indies.

The Family Muscidae includes house flies, face flies, stable flies, horn flies, and tsetse flies. The house fly, *Musca domestica*, is a cosmopolitan pest, and a mechanical vector of the etiologic agents of typhoid (*Salmonella typhosa*), yaws (*Treponema pertenue*), dysentery (*Entamoeba histolytica*) and cholera (*Vibrio comma*). The fly breeds in decaying organic materials. Adults feed upon a wide variety of foods, which they liquify with saliva, and then lap up with their proboscis. The larvae have been reported as causing accidental cutaneous, gastrointestinal and urinary myiasis in humans.

Other medically important species of *Musca* include the face flies, *M. autumnalis*, *M. vicina* and *M. sorbens*, which crawl upon the face and near the eyes, mouth and nares. Some species of *Musca* serve as intermediate hosts of *Thelazia* spp., the eye worms of cattle, dogs and humans. The larval nematodes escape from the proboscis of the fly when the insect feeds on or around the eyes of the vertebrate host. Some species are capable of transmitting conjunctivitis. The stable fly, *Stomoxys calcitrans*, is a ravenous bloodsucker that feeds on horses and other domestic animals. The fly frequently attacks humans, inflicting painful bites. Both sexes suck blood. The females oviposit in manure. The larvae develop in the manure and later pupate in the surrounding soil. Instances of cutaneous and intestinal myiasis have been reported. The repeated feeding of adults with contaminated mouthparts on different hosts may result in the mechanical transmission of trypanosomiasis and cutaneous Leishmaniasis.

The tsetse fly, *Glossina*, is the vector of the etiologic agents of African sleeping sickness. Both males and females feed on cattle, wild animals, and humans. Female flies give birth to fully developed larvae. The eggs, which are produced one at a time, hatch in utero. Special intrauterine glands ("milk glands") secrete nourishment for the larvae, which develop through three larval stages. Fourth-stage larvae are deposited in the soil where pupation occurs. The principal vector of Gambian trypanosomiasis is *G. palpalis*, while *G. morsitans* and *G. swynnertoni* are the major vectors of Rhodesian trypanosomiasis.

The Family Calliphoridae includes the blow flies, screw worm flies, and bluebottle flies. The adults are not bloodfeeders, but the larvae of several species cause myiasis in domestic animals and in humans (Fig. 12). Female screw worm flies deposit and glue their eggs in masses at the edges of open wounds. After the eggs hatch, the larvae ingest host tissues, penetrate the wound, invade the sinuses, eyes, ears, mouth, and genital passages, producing painful lesions which characteristically become secondarily infected. The species of major importance include *Callitroga hominivorax* (the primary screw worm), and *Chrysomya* spp. (generally a secondary invader of sores). The former species is widely distributed from the southern United States to northern Chile. The latter species occurs throughout southern Asia and in parts of Africa. The tumbu fly, *Cordylobia anthropophaga*, is distributed throughout much of Africa. Eggs are deposited on the ground, and the larvae actively penetrate the skin of humans or other animals in contact with contaminated soil. The larvae feed on host tissues and produce indurated, cutaneous swellings from which they eventually emerge, drop to the ground, enter the soil and pupate.

Flesh flies, which belong to the family Sarcophagidae, are parasitic only during the larval stages, and thus are capable of producing myiasis. Female sarcophagids are larviparous, depositing their young in masses on carrion, feces, open wounds, or on healthy, undamaged body surfaces. Myiasis-causing species include *Sarcophaga* spp., *Titanogrypha* spp., *Wohlfahrtia magnifica* and *W. vigil*. Only the latter species larviposits on healthy skin, which the larvae actively penetrate. The larvae of all species characteristically produce small raised lesions as they burrow in the subcutaneous tissues of their hosts. The lesions communicate with the surface through a small pore.

The Family Hippoboscidae is comprised of louse flies, which are permanent bloodsucking parasites of birds and mammals. Adult hippoboscids look neither like lice nor flies, but more like six-legged ticks. The sheep ked, *Melphagous ovinus*, is a wingless ectoparasite of sheep and goats. Both sexes are voracious bloodsuckers. The parasite is distributed worldwide except in the tropics. The entire life cycle is spent on the host. Gravid females larviposit approximately one larva per week over a period of four months. Almost immediately after deposition, the larvae attach to the host's wool and pupate. Heavy infestations cause severe irritation, emaciation and anemia. The insects readily attack sheep shearers and other persons associated with infested animals, inflicting painful bites. Another hippobosid that occasionally bites humans is *Pseudolynchia canariensis,* or pigeon fly. The parasite is common on pigeons throughout most of the warm, temperate areas of the world.

The names warble fly and bot fly frequently are used interchangeably for flies belonging to the families Cuterebridae, Gastrophilidae, Oestridae, and Hypodermatidae. Warble flies are so-called because their larvae live for a period of time just under the skin of their vertebrate hosts and produce a swelling or "warble". *Dermatobia hominis*, or the human bot fly, is a member of the family Cuterbridae that occurs in Central and South America. Unlike other myiasis-causing flies, *D.*

8

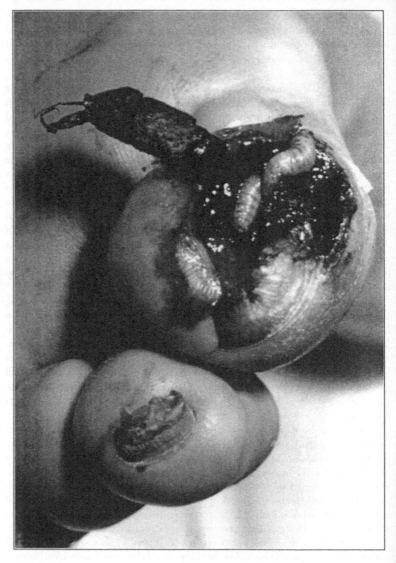

Figure 12. Dipterous larvae feeding on an open wound. Photograph courtesy of Dianora Niccolini.

hominis does not lay its eggs directly on its host. Instead, the non-biting female cements her eggs to the bodies of ticks, mosquitoes, or other dipterans, engaging these arthropods in transporting the eggs to humans or other suitable mammals. After the bot fly eggs hatch, the larvae penetrate the host's skin. The larvae migrate to the subcutaneous tissues and produce inflamed, boil-like pockets, each with a separate opening to the surface. After a period of one to two months, the mature larvae emerge from the skin, drop to the ground and pupate. In humans, cutaneous lesions frequently develop on the wrists, ankles, neck and face. Heavy infestations of cattle and dogs may terminate fatally. Early surgical removal of the larvae is an effective treatment. The family Gastrophilidae comprises the stomach bots of equids, elephants, and rhinoceroses. The adults, which are similar to honeybees in size and appearance, are nonfeeders. The parasitic larvae feed on the mucous of the host's stomach and intestines, causing enteric myiasis. Heavy infections cause severe damage to the host's gut, and if not treated early terminate fatally.

Three species are common in North America and throughout most of the world; *Gastrophilus intestinalis*, the horse bot, *G. nasalis*, or throat bot and *G. haemorrhoidalis*, the nose bot. Females of *G. intestinalis* attach their eggs to the hairs of the horse, usually on the knees and lower legs. When the horse licks these areas, the friction, warmth and moisture provided by the tongue stimulate hatching. The first-stage larvae penetrate the lips and tongue and tunnel down to the stomach and small intestine. When mature, the larvae detach from the mucosa of the alimentary canal, pass from the host in the feces and undergo pupation in the soil. The eggs of *G. nasalis* are usually attached to the hairs on the underside of the jaws, and the larvae crawl into the mouth where they feed for a time in pockets between the teeth before migrating to the gut. The eggs of *G. haemorrhoidalis* are deposited around the lips. The route of larval migration to the gut parallels that of *G. intestinalis* except that third-stage larvae attach to the anus for a short time before passing out with the feces. Occasionally, larvae of all three species penetrate human skin and mucous membranes, causing a creeping myiasis as they migrate under the epidermis. The serpiginous tunnels formed by these migrating fly larvae resemble cutaneous larva migrans produced by hookworm larvae. Human infestations usually occur on the extremities, and the maggots can be easily removed surgically.

Hypoderma bovis, the cattle bot, and *H. lineatum*, the heel fly, are two cosmopolitan species belonging to the family Hypodermatidae. The adult flies, which resemble bumble bees in size and appearance, do not feed. Females cement their eggs on the legs or body hairs of cattle. On hatching, the larvae penetrate the skin and, for a period of several months, migrate through the body, invading the intestines, liver, heart, esophagus, and other organs. Eventually, the larvae migrate to the back of the hosts, where they produce subcutaneous swellings termed warbles. The larvae make small holes in the skin through which they obtain air. The larvae remain in the skin for about two months, during which time they feed and grow to a length of about 25 mm. When mature, the larvae exit the skin and pupate in the soil. Cutaneous myiasis frequently develops in humans who are in contact with heavily infested cattle. The lesions, which are frequently inflamed and very painful, are usually on the extremities and in the neck region. Larvae of *Hypoderma*, unlike those of *Gastrophilus*, can successfully migrate and develop in humans.

Head maggots, which belong to the family Oestridae, are similar in size and shape to honeybees. The adult flies do not feed, but the larvae invade and parasitize the nasal cavities and sinuses of sheep, goats, horses, and other hoofed animals. The sheep bot, *Oestrus ovis*, is a cosmopolitan parasite of domestic sheep and goats. The female is larviparous and deposits maggots in the nares of their hosts. The larvae then enter the sinuses and attach to the mucous membranes and feed. The mature

8

larvae emerge from the nostrils and drop to the ground to pupate in the soil. *Rhinostrus purpureus* is a head maggot of horses in Europe, northern Africa, parts of Asia and Central and South America. Females larviposit onto the eyes and nostrils of the host, and occasionally cause opthalmic myiasis. The larvae, which do not mature in humans, cause inflammation, conjunctivitis, and in severe cases blindness.

SELECTED REFERENCES

The following references include relatively recent articles of a general nature that address varied topics concerned with tropical medicine, hygiene, and host-parasite relationships.

1. Abath FGC. Development of vaccines against human parasitic diseases: Tools, current status and perspectives. Expert Opinion on Investigational Drugs 2000; 9:301-310.

2. Abath FGC, Montenegro SML, Gomes YM. Vaccines against human parasitic diseases: An overview. Acta Tropica 1998; 71:237-254.

3. Ambroise-Thomas P. Emerging parasite zoonoses: the role of host-parasite relationship. Internat J Parasitol 2000; 30:1361-1367.

4. Ambroise-Thomas P. Parasitic diseases and immunodeficiencies. Parasitology, 2001; 122:S65-S71.

5. Arya SC. Global warming and the performance of drugs used to treat parasitic and other diseases. Ann Trop Med Parasitol 1999; 93:207-208.

6. Ashford RW. Current usage of nomenclature for parasitic diseases, with special reference to those involving arthropods. Med Vet Entomol 2001; 15:121-125.

7. Aspock H, Auer H, Picher O. Parasites and parasitic diseases in prehistoric human populations in Central Europe. Helminthologia 1999; 36:139-145.

8. Ayles HM, Corbett EL, Taylor I, Cowie AGA, Bligh J, Walmsley K et al. A combined medical and surgical approach to hydatid disease: 12 years' experience at the Hospital for Tropical Diseases, London. Ann Royal Coll Surg Eng 2002; 84:100-105.

9. Baker DG, Bryant JD, Urban JF, Lunney JK. Swine immunity to selected parasites. Vet Immunol Immunopathol 1994; 43:127-133.

10. Barcinski MA, Costamoreira ME. Cellular response of protozoan parasites to host-derived cytokines. Parasitol Today 1994; 10:352-355.

11. Barcinski MA, DosReis GA. Apoptosis in parasites and parasite-induced apoptosis in the host immune system: A new approach to parasitic diseases. Brazilian J Med Biolog Res 1999; 32:395-401.

12. Barillas-Mury C, Wizel B, Han YS. Mosquito immune responses and malaria transmission: lessons from insect model systems and implications for vertebrate innate immunity and vaccine development. Insect Biochem Mol Biol 2000; 30:429-442.

13. Barry M, Maguire JH, Weller PE. The American Society of Tropical Medicine and Hygiene initiative to stimulate educational programs to enhance medical expertise in tropical diseases. Am J Trop Med Hyg 1999; 61:681-688.

14. Beerntsen BT, James AA, Christensen BM. Genetics of mosquito vector competence. Microbiol Mol Biol Rev 2000; 64:115-+.

15. Bell A. Microtubule inhibitors as potential antimalarial agents. Parasitol Today 1998; 14:234-240.

16. Bergquist NR. Vector-borne parasitic diseases: New trends in data collection and risk assessment. Acta Tropica 2001; 79:13-20.

17. Blancou J, Meslin FX. A brief historical overview of zoonoses. Revue Scientifique Et Technique De L Office International Des Epizooties 2000; 19:15-22.

18. Borst P, Ouellette M. New mechanisms of drug resistance in parasitic protozoa. Ann Rev Microbiol 1995; 49:427-460.

19.   Burkhart CN. Ivermectin: An assessment of its pharmacology, microbiology and safety. Vet Hum Toxicol 2000; 42:30-35.

20.   Bynum WF. Transmissible diseases at the end of the XIXth century: Vectors and pathogenic agents. Bulletin de la Societe de Pathologie Exotique 1999; 92:408-410.

21.   Carlier Y, Truyens C. Influence of maternal infection on offspring resistance towards parasites. Parasitol Today 1995; 11:94-99.

22.   Chakraborty AK, Majumder HK. Molecular biology of *Leishmania*: Kinetoplast DNA and DNA topoisomerases as novel therapeutic targets. Current Science 1999; 76:1462-1472.

23.   Chang KH, Han MH. MRI of CNS parasitic diseases. J Mag Reson Imag 1998; 8:297-307.

24.   Christensen BM, Ferdig MT. The field of vector biology—A perspective on the first 100 years. Parasitol Today 1997; 13:295-297.

25.   Chu DTW. The future role of quinolones. Expert Opinion on Therapeutic Patents 1996; 6:711-737.

26.   Colley DG. Parasitic diseases: Opportunities and challenges in the 21st century. Memorias do Instituto Oswaldo Cruz 2000; 95:79-87.

27.   Combes C. Fitness of parasites—Pathology and selection. Internat J Parasitol 1997; 27:1-10.

28.   Coop RL, Kyriazakis I. Nutrition-parasite interaction. Vet Parasitol1999; 84:187-204.

29.   Couturier E, Hansmann Y, Descampeaux C, Christmann D. The epidemiology of infective endocarditis. Medecine et Maladies Infectieuses 2000; 30:162-168.

30.   Dailey MD. Parasitic diseases. CRC Handbook Of Marine Mammal Medicine. Second Edition. 2001.

31.   Dedet JP. Update on leishmaniases. Presse Medicale 2000; 29:1019-1026.

32.   Diouf S, Diagne I, Moreira C, Sy HS, Faye O, Ndiaye O et al. Iron, vitamin A deficiencies and parasitic diseases: Impact on Senegalese children developement. Archives de Pediatrie 2002; 9:102-103.

33.   Dodd RY. Transmission of parasites by blood transfusion. Vox Sanguinis 1998; 74:161-163.

34.   Duvic C, Nedelec G, Debord T, Herody M, Didelot F. Imported parasitic nephritis: Review of medical literature. Nephrologie 1999; 20:65-74.

35.   Ebert D, Hamilton WD. Sex against virulence—The coevolution of parasitic diseases. Trends Ecol Evol 1996; 11:A 79-A 82.

36.   Ebert D, Herre EA. The evolution of parasitic diseases. Parasitol Today 1996; 12:96-101.

37.   Edwards G, Winstanley PA, Ward SA. Clinical pharmacokinetics in the treatment of tropical diseases—Some applications and limitations. Clinical Pharmacokinetics 1994; 27:150-165.

38.   Eyckmans L. Teaching tropical medicine—A personal view in perspective. Ann Trop Med Parasitol 1997; 91:743-746.

39.   Feldmeier H, Poggensee G, Krantz I, Hellinggiese G. Female genital schistosomiasis—New challenges from a gender perspective. Trop Geograph Med 1995; 47:S 2-S 15.

40.   Gallup JL, Sachs JD. The economic burden of malaria. Am J Trop Med Hyg 2001; 64:85-96.

41.   Giamarellou H. AIDS and the skin: Parasitic diseases. Clinics in Dermatology 2000; 18:433-439.

42.   Gilles HM, Lucas AO. Tropical medicine—100 years of progress. Br Med Bull 1998; 54:269-280.

43. Gomez-Lus R, Clavel A, Castillo J, Seral C, Rubio C. Emerging and reemerging pathogens. Internat J Antimicrob Agents 2000; 16:335-339.

44. Gubler DJ. Prevention and control of tropical diseases in the 21st century: Back to the field. Am J Trop Med Hyg 2001; 65:V-XI.

45. Harms G, Feldmeier H. Review: HIV infection and tropical parasitic diseases—Deleterious interactions in both directions? Trop Med International Health 2002; 7:479-488.

46. Hellinggiese G, Feldmeier H, Racz P, Hickl EJ. Female genital schistosomiasis (FGS)—Literature review and two case presentations. Geburtshilfe und Frauenheilkunde 1997; 57:136-140.

47. Heyworth MF. Parasitic diseases in immunocompromised hosts—Cryptosporidiosis, isosporiasis, and strongyloidiasis. Gastroenterology Clinics of North America 1996; 25:691 ff.

48. Hoberg EP, Alkire NL, de Queiroz A, Jones A. Out of Africa: Origins of the Taenia tapeworms in humans. Proceedings of the Royal Society of London—Series B: Biological Sciences 2001; 268:781-787.

49. Houszka M. Metastrongylosis as an agent in the population decrease of wild boars. Medycyna Weterynaryjna 2001; 57:638-640.

50. Hutubessy RCW, Bendib LM, Evans DB. Critical issues in the economic evaluation of interventions against communicable diseases. Acta Tropica 2001; 78:191-206.

51. Ijumba JN, Lindsay SW. Impact of irrigation on malaria in Africa: Paddies paradox. Med Vet Entomol 2001; 15:1-11.

52. Innes EA. Emerging parasitic diseases, bioterrorism and the New World order. Parasitol Today 1999; 15:427-428.

53. Irwin PJ. Companion animal parasitology: A clinical perspective. Intern J Parasitol 2002; 32:581-593.

54. Jacobson ER. Causes of mortality and diseases in tortoises—A review. J Zoo Wildlife Med 1994; 25:2-17.

55. James WD. Imported skin diseases in dermatology. J Dermatol 2001; 28:663-666.

56. Johnston DA, Blaxter ML, Degrave WM, Foster J, Ivens AC, Melville SE. Genomics and the biology of parasites. Bioessays 1999; 21:131-147.

57. Jokiranta TS, Jokipii L, Meri S. Complement resistance of parasites. Scand J Immunol 1995; 42:9-20.

58. Kaplan JE, Hu DJ, Holmes KK, Jaffe HW, Masur H, Decock KM. Preventing opportunistic infections in human immunodeficiency virus-infected persons—Implications for the developing world. Am J Trop Med Hyg 1996; 55:1-11.

59. Kimber KR, Kollias GV. Infectious and parasitic diseases and contaminant-related problems of North American river otters (Lontra canadensis): A review. J Zoo Wildlife Med 2000; 31:452-472.

60. Kinnamon KE, Engle RR, Poon BT, Ellis WY, McCall JW, Dzimianski MT. Polyamines: Agents with macrofilaricidal activity. Ann Trop Med Parasitol 1999; 93:851-858.

61. Kristensen TK, Brown DS. Control of intermediate host snails for parasitic diseases—A threat to biodiversity in African freshwaters? Malacologia 1999; 41:379-391.

62. Le Souef PN, Goldblatt J, Lynch NR. Evolutionary adaptation of inflammatory immune responses in human beings. Lancet 2000; 356:242-244.

63. Liance M. Progress toward the development of vaccines against parasitic diseases—A review based on studies performed by participants of the laveran seminar. Parasite-Journal de la Societe Francaise de Parasitologie 1994; 1:197-203.

64. Lightowlers MW, Gauci CG. Vaccines against cysticercosis and hydatidosis. Vet Parasitol 2001; 101:337-352.

65. Londershausen M. Approaches to new parasiticides. Pesticide Science 1996; 48:269-292.
66. Luder CGK, Gross U, Lopes MF. Intracellular protozoan parasites and apoptosis: Diverse strategies to modulate parasite-host interactions. Trends Parasitol 2001; 17:480-486.
67. Mackenzie CD. Anthelmintic therapy—Current approaches and challenges. Current Opinion in Infectious Diseases 1993; 6:812-823.
68. Mahe A. Bacterial skin infections in a tropical environment. Curr Opin Infect Dis 2001; 14:123-126.
69. Maizels RM, Holland MJ. Parasite immunity—Pathways for expelling intestinal helminths. Current Biology 1998; 8:R 711-R 714.
70. Marton LJ, Pegg AE. Polyamines as targets for therapeutic intervention. Ann Rev Pharmacol Toxicol 1995; 35:55-91.
71. Michael E, Grenfell BT, Isham VS, Denham DA, Bundy DAP. Modelling variability in lymphatic filariasis—Macrofilarial dynamics in the *Brugia pahangi* cat model. Proceedings of the Royal Society of London—Series B: Biological Sciences 1998; 265:155-165.
72. Mitchell G, Nossal G. Funding options for research: Facing the market as well as government. Internat J Parasitol 1999; 29:819-831.
73. Molyneux DH. Vectorborne parasitic diseases—An overview of recent changes. Internat J Parasitol 1998; 28:927-934.
74. Morgado MG, Barcellos C, Pina MD, Bastos FI. Human immunodeficiency virus/ acquired immunodeficiency syndrome and tropical diseases: A Brazilian perspective. Memorias do Instituto Oswaldo Cruz 2000; 95:145-151.
75. Mott KE, Nuttall I, Desjeux P, Cattano P. New geographical approaches to control of some parasitic zoonoses. Bulletin of the World Health Organization 1995; 73:247-257.
76. Murray M. The parasites, predators, places and people I have known: A great adventure. Vet Parasitol 1999; 81:149-158.
77. Nozais JP. Parasitic diseases and fecal peril: Diseases due to helminths. Bulletin de la Societe de Pathologie Exotique 1998; 91:416-422.
78. Osman M, Lausten SB, Elsefi T, Boghdadi I, Rashed MY, Jensen SL. Biliary parasites. Digestive Surgery 1998; 15:287-296.
79. Ouaissi A, Ouaissi M, Sereno D. Glutathione S-transferases and related proteins from pathogenic human parasites behave as immunomodulatory factors. Immunol Lett 2002; 81:159-164.
80. Pantanowitz L, Telford SR, Cannon ME. Tick-borne diseases in transfusion medicine. Transfusion Medicine 2002; 12:85-106.
81. Papadopoulou B, Kundig C, Singh A, Ouellette M. Drug resistance in leishmania— Similarities and differences to other organisms. Drug Resistance Updates 1998; 1:266-278.
82. Papadopuloseleopulos E, Turner VF, Papadimitriou JM, Bialy H. AIDS in Africa— Distinguishing fact and fiction. World J Microbiol Biotechnol 1995; 11:135-143.
83. Patel R, Paya CV. Infections in solid-organ transplant recipients. Clinical Microbiology Reviews 1997; 10:86 ff.
84. Patz JA, Graczyk TK, Geller N, Vittor AY. Effects of environmental change on emerging parasitic diseases. Internat J Parasitol 2000; 30:1395-1405.
85. Peng SL. Rheumatic manifestations of parasitic diseases. Seminars in Arthritis & Rheumatism 2002; 31:228-247.
86. Perlmann P, Bjorkman A. Malaria research: host-parasite interactions and new developments in chemotherapy, immunology and vaccinology. Curr Opin Infect Dis 2000; 13:431-443.

87. Pointier JP. Invading freshwater gastropods: Some conflicting aspects for public health. Malacologia 1999; 41:403-411.

88. Poli G, Vicenzi E, Ghezzi S, Lazzarin A. Cytokines in the acquired immunodeficiency syndrome and other infectious diseases. Internat J Clin Lab Res 1995; 25:128-134.

89. Repetto R, Baliga SS. Pesticides and immunosuppression—The risks to public health. Health Policy & Planning 1997; 12:97-106.

90. Roberts CW, Walker W, Alexander J. Sex-associated hormones and immunity to protozoan parasites. Clin Microbiol Rev 2001; 14:476-+.

91. Roberts T, Murrell KD, Marks S. Economic losses caused by foodborne parasitic diseases. Parasitol Today 1994; 10:419-423.

92. Rodriguez JB. Specific molecular targets to control tropical diseases. Curr Pharm Des 2001; 7:1105-1116.

93. Ruppel A, Doenhoff MJ. Vector biology and the control of parasitic diseases. Parasitol Today 1998; 14:299-300.

94. Ryan TJ. Women in dermatology—Gender and tropical diseases. Internat J Dermatol 1995; 34:226-235.

95. Sangster NC, Gill J. Pharmacology of anthelmintic resistance. Parasitol Today 1999; 15:141-146.

96. Sansom C. New drugs needed for tropical diseases. Lancet Infect Dis 2002; 2:134.

97. Scully C, Monteil R, Sposto MR. Infectious and tropical diseases affecting the human mouth. Periodontology 2000; 18:47-70.

98. Sepulvedaboza S, Cassels BK. Plant metabolites active against *Trypanosoma cruzi*. Planta Medica 1996; 62:98-105.

99. Stanley SL. Protective immunity to amebiasis: New insights and new challenges. J Infect Dis 2001; 184:504-506.

100. Stanley SL, Virgin HW. *Scid* mice as models for parasitic infections. Parasitol Today 1993; 9:406-411.

101. Szenasi Z, Endo T, Yagita K, Nagy B. Isolation, identification and increasing importance of free-living amoebae causing human disease. J Med Microbiol 1998; 47:5-16.

102. Taverne J. Transgenic mice and the study of cytokine function in infection. Parasitol Today 1994; 10:258-262.

103. Thomas PK. Tropical neuropathies. J Neurol 1997; 244:475-482.

104. Thompson RCA. The future of veterinary parasitology: a time for change? Vet Parasitol 2001; 98:41-50.

105. Trouiller P, Olliaro P, Torreele E, Orbinski J, Laing R, Ford N. Drug development for neglected diseases: A deficient market and a public-health policy failure. Lancet 2002; 359:2188-2194.

106. Trouiller P, Olliaro PL. Drug development output: What proportion for tropical diseases? Lancet 1999; 354:164.

107. Uilenberg G. Integrated control of tropical animal parasitoses. Trop An Health Prod 1996; 28:257-265.

108. Ullman B, Carter D. Hypoxanthine-guanine phosphoribosyltransferase as a therapeutic target in protozoal infections. Infectious Agents & Disease 1995; 4:29-40.

109. Valerio L, Sabria M, Fabregat A. Tropical diseases in the Western world. Medicina Clinica 2002; 118:508-514.

110. Vanhamme L, Pays E. Control of gene expression in trypanosomes. Microbiologic Rev 1995; 59:223-240.

111.   Vermeulen AN. Progress in recombinant vaccine development against coccidiosis—A review and prospects into the next millennium. Internat J Parasitol 1998; 28:1121-1130.

112.   Vial HJ, Traore M, Fairlamb AH, Ridley RG. Renewed strategies for drug development against parasitic diseases. Parasitol Today 1999; 15:393-394.

113.   Vuitton DA, Godot V, Harraga S, Liance M, Beurton I, Bresson-Hadni S. Echinococcoses: A parasitic model for understanding allergic diseases? Revue Francaise d Allergologie et d Immunologie Clinique 2001; 41:285-293.

114.   Wasson K, Peper RL. Mammalian microsporidiosis. Vet Pathol 2000; 37:113-128.

115.   Werbovetz KA. Target-based drug discovery for malaria, leishmaniasis, and trypanosomiasis. Current Medicinal Chemistry 2000; 7:835-860.

116.   Zarnbrano-Villa S, Rosales-Borjas D, Carrero JC, Ortiz-Ortiz L. How protozoan parasites evade the immune response. Trends Parasitol 2002; 18:272-278.

117.   Zingales B, Rondinelli E, Degrave W, Dasilveira JF, Levin M, Lepaslier D et al. The *Trypanosoma cruzi* genome initiative. Parasitol Today 1997; 13:16-22.

GLOSSARY

Acanthella
The larval stage of an acanthocephalan following the acanthor and prior to the cystacanth.

Acanthor
The larval stage of an acanthocephalan; that develops inside an egg capsule and possesses bladelike hooks.

Accidental or incidental host
An organism, other than the normal host species, in which a parasite may or may not continue its development.

Acetabulum
Ventral sucker or holdfast structure of digenetic trematodes.

Acoelomate
The absence of a body cavity. The internal organs (viscera) lie embedded in parenchyma.

Acquired immunity
A host's immune response to a previous infection.

Ala (pl. alae)
Cuticular wing-like projection in certain nematodes.

Allergy
See hypersensitivity.

Alveolus (pl. alveoli)
An air sac in the lung where gaseous exchange occurs.

Amastigote
A small, ovoid, form of hemoflagellates that develops intracellularly.

Ametabolous
A type of insect metamorphosis in which there is no significant morphological change as the organism proceeds through a series of molts to the adult stage.

Amphids
Sensory depressions or pits, believed to be chemoreceptors, located anteriorly on the body surface of certain nematodes.

Anaphylaxis (anaphylactic shock)
An exaggerated or hypersensitive reaction by a host in response to a foreign protein (allergen) involving histamine-release.

Anapolysis
The process in which terminal, gravid proglottids or segments of a tapeworm are not shed.

Anisogamete
Morphologically different male and female gametes.

Anorexia
Loss of appetite.

Anterior station
The development of protozoan parasites in the anterior part of an insect vector, with transmission to the definitive host occurring when the vector bites.

Anthelmintic
A chemical used to remove worms, usually from the intestinal tract.
Anthrax
A bacterial disease of humans, cattle, and sheep, transmitted by tabanid flies.
Antibody
A specific serum protein (immunoglobulin) synthesized by B lymphoid cells (plasma cells) in response to an antigen.
Antigen
A substance, usually a protein, capable of inducing the host to synthesize antibodies.
Antigenic mimicry
Acquisition or production of host-like molecules by a parasite so that it is not recognized by the host as foreign, thus circumventing an immune response.
Apical complex
A combination of secretary structures found in the apical region of sporozoites and merozoites of members of the phylum Apicomplexa.
Apicomplexa
A phylum containing animals whose life cycles include feeding stages (trophozoites), asexual multiplication (schizogony), and sexual multiplication (gametogony and sporogony).
Apolysis
The process in which terminal, gravid or egg-filled proglottids are detached and shed from tapeworm strobila.
Arachnida
A class in the phylum Arthropoda containing ticks, mites, spiders, and scorpions.
Arbovirus
A virus transmitted from one human to another by an arthropod.
Arthropoda
A phylum comprised of animals having a chitinous exoskeleton and paired jointed legs. Includes insects, crustaceans and arachnids.
Atrium (pl. atria)
An opening into the body (e.g., mouth, urethra).
Autogeny
The ability of some bloodsucking arthropods to lay eggs without having had a blood meal (adj. autogenous).
Autoinfection
Reinfection of a host by the progeny of an existing parasitic organism residing within the host.
Axoneme
An intracellular microtubular portion of the flagellum or cilium.
Axostyle
A tube-shaped sheath of microtubules, observed in many flagellates (e.g., trichomonads), that usually extends from a basal body to the posterior end, and functions as a supporting cytoskeleton.
B cell
A specialized lymphocyte that produces antibodies.
Basal body (= Blepharoplast)
A centriole-like organelle from which the flagellum or cilium originates.

Biological vector
A host that is required for the development of the parasite and for transferring the parasite to another host.

Biramous
A structure divided into two branches.

Blackwater fever
Massive lysis of vertebrate erythrocytes that occasionally accompanies falciparum malaria.

Bothridium (pl. bothridia)
A muscular, leaf-like adhesive groove on the scolex of certain tapeworms of the order Pseudophyllidea.

Bradyzoites
Slowly multiplying intracellular trophozoites of *Toxoplasma gondii* that form pseudocysts in immune hosts.

Brood capsule
A structure within the daughter cyst in Echinococcus granulosus, in which numerous scolices develop. In the definitive host each scolex can develop into an adult tapeworm.

Buccal capsule (cavity)
The oral cavity of roundworms; it may contain either teeth or cutting plates.

Bug
An insect of the order Hemiptera.

Bursa (pl. bursae)
A muscular copulatory structure.

Calabar swelling
A transient, subcutaneous swelling caused by the nematode *Loa loa*.

Capitulum
A collective term referring to the mouthparts of ticks and mites extending forward from the head of the arthropod.

Carrier
A host harboring and disseminating a parasite but exhibiting no clinical symptoms.

Cell mediated reaction
The effect produced by specialized T lymphocytes.

Cellular immunity
A specific response to an antigen in which lymphoid cells are the primary effectors.

Cercocystis
A modified cysticercoid larva of *Hymenolepis nana* found in the intestinal villus of the definitive host.

Cercomer
Tail-like appendage on tapeworm procercoid and cysticercoid larvae frequently possessing hooks of the hexacanth embryo.

Cestoda
A class within the phylum Platyhelminthes comprised of tapeworms.

Chagas' disease
A disease, also known as American trypanosomiasis, caused by the flagellate *Trypanosoma cruzi*.

Chigger
A mite of the family Trombiculidae.

Chitin
An insoluble polysaccharide found in the exoskeletons of arthropods.
Chromatoidal body (or bar)
A rod-shaped structure in the cytoplasm of some ameba cysts.
Cilia
Hair-like processes attached to the cell surface and used for motility through fluids.
Ciliophora
A phylum containing animals that move by means of cilia and that have two dissimilar nuclei.
Cirrus
The penis or ejaculatory duct of a flatworm.
Coenurus
A larval tapeworm comprised of numerous scolices that bud from internal germinal epithelia.
Commensalism
The association of two different species of organisms in which one partner is benefited and the other is neither benefited nor injured. A type of symbiosis in which there is no discernible damage to the host.
Conjugation
A temporary union of two ciliated protozoans for the exchange of nuclear material.
Copulatory spicules
Needlelike structures possessed by some male nematodes used during copulation.
Coracidium
Free-swimming, ciliated embryophore of tapeworms of the order Pseudophyllidea.
Costa
A striated, rod-like structure that lies just under the recurrent flagellum of certain protozoa of the order Trichomonadida.
Creeping eruption
The irritation and rash caused by the migration under the skin of non-human hookworm larvae.
Crustacea
A class in the phylum Arthropoda that is comprised of crabs, water fleas, lobsters, shrimp, and barnacles.
Cryptozoite
The exoerythrocytic stage stage in the life cycle of *Plasmodium* spp. developing in liver cells.
Ctenidium (pl. ctenidia)
Comb-like structures found on the head region of fleas. Genal combs are located just above the mouthparts. Pronotal combs are located immediately behind the head and extend posteriority on the dorsal surface.
Cutaneous larva migrans
A disease caused by the migration under the skin of humans of larvae of *Ancylostoma* spp. (frequently dog or cat hookworms) or other roundworms. Cutaneous larval migration, which also is termed creeping eruption, is marked by thin, red, papular lines of eruption on the skin.
Cuticle
The outer protective surface of helminths and arthropods.

Cyst
A general term used to describe the resistant stage in the life cycle of an organism. This stage is frequently infective to a new host.

Cysticercoid
A tapeworm larva in which the non-inverted scolex is surrounded by a fluid-filled bladder. This stage is characteristic of tapeworms belonging to the families Hymenolepididae, Dilepidiae, or Anoplocephalidae.

Cysticercosis
Infection with cysticercus larvae.

Cysticercus
A tapeworm larva in which the inverted scolex is surrounded by a fluid-filled bladder. This stage is characteristic of the cyclophyllidean family Taeniidae.

Cystogenous glands
Secretory cells in the cercariae of some digenetic Trematodes that give rise to metacercarial cysts.

Cytostome
Rudimentary mouth.

Definitive host
The animal in which a parasite passes its adult existence and/or sexual reproductive phase.

Delayed hypersensitivity
Increased reactivity to a specific antigen, mediated by cells rather than antibodies, usually requiring up to 24 hours to reach maximum intensity.

Dengue (= blackwater fever)
A disease caused by a mosquito-transmitted virus.

Dermatitis
Inflammation of the skin.

Diagnostic stage
The developmental stage(s) of a pathogenic that aids in its identification.

Differential diagnosis
The clinical comparison of different diseases that exhibit similar symptoms designed to determine from which the patient is suffering.

Direct life cycle
A life cycle in which only a single host is required for the successful development of a parasite or pathogen.

Diurnal
Occurring during the daytime.

Dysentery
A form of diarrhea involving the discharge in the feces of blood and mucus.

Ecdysis
The molting or shedding of an outer layer or covering (cuticle) and the development of a new one.

Ectoparasite
A parasite that lives on the exterior surface or in the integument of a host.

Ectopic site
Abnormal or unexpected site of infection.

Endoparasite
A parasite established within the body of its host.
Edema
Swelling resulting from the abnormal accumulation of fluid in cells, tissues, or tissue spaces.
Elephantiasis
Overgrowth of the skin and subcutaneous tissue due to obstructed circulation of lymph in the lymphatic vessels caused by the roundworm *Wuchereria bancrofti*.
Embryonation
The development of a fertilized helminth embryo into a larva.
Embryophore
The shell of *Taenia* and other tapeworm eggs.
Endemic
A disease or disease agent that occurs in a human community at all times.
Endodyogeny
A special form of merogony in which two daughter cells are while still in the mother cell. This processes occurs in certain members of the phylum Apicomplexa (e.g., Toxoplasma).
Endoparasite
A parasite that lives inside the host.
Endosome
The small mass of chromatin within the nucleus, comparable to a nucleolus of metazoan cells (also termed karyosome).
Endosome
Inflammation of the intestine.
Entomology
The branch of zoology dealing with the study of insects.
Enzootic
A disease or disease agent that persists in an animal population.
Eosinophilia
A disease manifested by high levels of blood eosinophils.
Epidemic
A disease or disease agent that spreads rapidly through a human population.
Epidemiology
The study of the relationship of the various factors that determine the frequency and distribution of an infectious process or disease in a community.
Epimastigote
A flattened, spindle-shaped form of a hemoflagellate, possessing a short undulating membrane, and a kinetoplast that lies anterior to the nucleus. Epimastigotes are seen primarily in the gut (e.g., in the reduviid bug) or salivary glands (e.g., in the tsetse fly) of the vectors in the life cycle of trypanosomes.
Epizootic
A disease or disease agent that spreads rapidly through an animal population.
Espundia
A disease caused by *Leishmania braziliensis*, also known as mucocutaneous Leishmaniasis, uta, pian bois, and chiclero ulcer.

Excystation
Transformation from a cyst to a trophozoite after the cystic form has been swallowed by the host.

Exflagellation
The process whereby a sporozoan microgametocyte releases haploid flagellated microgametes that can fertilize the macrogamete and thus form a diploid zygote, such as *Plasmodium*.

Exoskeleton
A hard, chitinous structure on the outside of the body, providing support for internal organs.

Feral
Pertains to parasites that occurs in the wild as opposed to an urban sites.

Filariform larva
Infective, non-feeding, sheathed, third-staged larva of nematodes.

Flagellum (pl. flagella)
An extension of ectoplasm that serves in locomotion.

Flame cell (=protonephridium)
The terminal cell of the excretory system in platyhelminths containing a group of luminal cilia that moves fluid through the tubule.

Gamogony (=gametogony)
Formation of gametes.

Gamete
A mature sex cell.

Gametocyte
The malaria sexual cell in human blood. Gametocytes produce gametes on the mosquito's stomach.

Gametogony
The phase of the development cycle of the malarial parasite in the human in which male and female gametocytes are formed.

Gastrodermis
The tissue lining the digestive tract, as found in digenetic trematodes.

Genital atrium (=genital pore)
A area in the body into which open male and female genital ducts.

Genus (pl. genera)
A taxonomic category subordinate to Family and superior to Species.

Granuloma
A swelling composed of cells (leukocytes), fluid, and connective tissue, frequently representing a host reaction against a foreign-body.

Gravid
A pregnant individual having developing eggs, embryos, or larvae.

Ground itch
Skin penetration by non-human hookworm larvae causing localized irritation and rash.

Gynecophoral canal
Longitudinal groove on the ventral surface of male schistosomes in which the female worm lies.

Helminth
A term for worms.

Hematophagous
Bloodsucking; usually refers to the feeding habits of various insects and ticks.
Hematuria
Blood in the urine.
Hemimetabolous
A type of insect development in which there is a gradual change in the external structure as development proceeds to the adult stage.
Hemocoel
The body cavity of arthropods, typically containing blood or hemolymph.
Hemozoin
Granules, seen in erythrocytes infected with Plasmodium malariae.
Hermaphroditic
Having both male and female reproductive organs within the same individual.
Heterogonic
The term used to describe a life cycle in which free-living generations may alternate periodically with parasitic generations. Reproduction in which sexual and asexual generations alternate, as in the nematode *Strongyloides.*
Heteroxenous
Requiring more than one host to complete a life cycle.
Hexacanth embryo
A tapeworm larva having six hooklets (see onchosphere).
Holometabolous
A type of insect development in which there are distinct morphological changes as the insect develops through one or more larval stages, a pupal stage, and the imago or adult stage.
Homogonic
The term used to describe a lifestyle that is consistently either parasitic or free-living.
Homoxenous
Refers to a parasite that has a direct life cycle, or one that requires only a single host.
Host specificity
The extent to which a parasite can exist in more than one host species.
Host, definitive
The host in which a parasite attains sexual maturity.
Host, intermediate
The host in which a parasite undergoes developmental changes but does not become sexually mature.
Host, paratenic
A transfer host in which a parasite resides without developing.
Host, reservoir
Usually a non-human host in which a parasite lives and remains a source of infection but usually shows no symptoms.
Host
An animal or plant that harbors a parasite, providing the latter with some metabolic resource.

Humoral immunity
A specific host response to an antigen in which the principal effectors are circulating antibodies that immobilize or destroy the antigen.

Hydatid
A larval form of tapeworm in which numerous scolices of potential tapeworms bud from secondary cysts. This larval form is characteristic of tapeworms in the genus *Echinococcus*.

Hydatid sand
Granular material consisting of free scolices, hooklets, and daughter cysts in the fluid of hydatid cysts of Echinococcus granulosus.

Hyper-
A prefix meaning more than normal, over or above.

Hyperemia
An abnormally large amount of blood in a tissue.

Hyperplasia
An abnormally high number of cells in a tissue.

Hypersensitivity (=allergy)
A condition in which an organism is sensitized to a particular substance and manifests an abnormally strong reaction upon subsequent exposures to the substance.

Hypertrophy
An abnormal increase in the size of a tissue.

Hypodermis
The tissue that secretes the overlying cuticle.

Imago
The sexually mature adult insect.

Immunity
A specific response, cellular and/or humoral, to a foreign protein (antigen).

Immunopathology
An immune response that is damaging in itself.

Immunosuppression
Depressed immune responsiveness.

In vitro
Observable in a test tube, or other non-living system.

In vivo
Within the body.

Incidence (of infection)
The proportion of a population infected or showing disease over a given period of time.

Incidental parasite
Parasite found in a host other than its usual one.

Incubation period
The interval of time from initial infection to the onset of clinical symptoms of a disease.

Indirect life cycle
A life cycle in which more than one host is required for the parasite or pathogen to complete its development.

Infection
Invasion of the body by a parasite or pathogen.

Infective stage
The stage of a parasite that is capable of entering the host and continuing development within the host.

Infestation
The establishment of a parasite on the external surface of a host.

Inflammation
A response to a physical, chemical or biological insult causing pain, reddening, increased temperature, and swelling at the site of injury.

Instar
A stage in the life cycle of an insect, such as a larval or nymphal instar.

Intermediate host
An animal that serves as host for only the larval or sexually immature stages of a parasite. An intermediate host is required for the successful development of the parasite.

Invertebrates
Animals having no spinal column.

Juvenile
A sexually immature stage of an organism.

Kala-azar
A disease, also known as visceral Leishmaniasis or dum-dum fever, caused by *Leishmania donovani*.

Kinetoplast
A mitochondrial-like organelle characteristic of protozoa of the order Kinetoplastida.

Kinetosome
See basal body.

L.D. body (Leishman-Donovan body)
Each of the small ovoid ametabolous forms found in macrophages of the liver and spleen in patients with *Leishmania donovani* infection.

Larva (pl. larvae)
An immature or pre-adult stage in the development of certain insects or nematodes.

Laurer's canal
A canal, originating on the surface of the oviduct near the seminal receptacle in some digenetic trematodes, that may represent a vestigial vagina.

Leishmaniasis
A disease caused by members of the genus *Leishmania*.

Lumen (pl. lumina)
The central cavity of an organ.

Lycophore
The ten-hooked or decacanth larva that develops within the egg capsules of tapeworms of the subclass Cestodaria.

Lymphokine
Any of several chemical mediators, released by T cells, that react with other cells essential to the inflammatory process.

Macrogametocyte
The cell that gives rise to a macrogamete.
Malpighian tubules
Excretory organs of terrestrial insects and most acarines.
Mastigophora
A subphylum of Protozoa containing organisms that possess one or more flagella.
Maurer's dots
Aggregates in cytoplasm of erythrocytes infected with Plasmodium falciparum.
Mechanical vector
A host that is not necessary for the development of a parasite, but one that serves as a temporary refuge for transfer of the parasite to an obligatory host.
Merogony
A type of asexual reproduction in which there is nuclear replication without plasmotomy, and then two to many daughter cells (i.e., merozoites) are produced simultaneously.
Merozoite
Usually an elongate stage in the life cycle of sporozoans produced by merogony that infects host cells to undergo additional merogony or to ultimately form gametes by gamogony.
Metacercaria
The larval stage between cercaria and adult in the life cycle of many digenetic trematodes.
Metacestode
Tapeworm stage following the oncosphere, but one not yet sexually mature.
Metacystic trophozoite
A small trophozoite of Entamoeba spp. that emerges from the cyst in the intestine of the host.
Metamerism
Segmental repetition of homologous parts; in each metamere or segment there are identical structures such as muscles, neural ganglia, and nephridia.
Metamorphosis
A change of shape or structure involving the transition from one developmental stage to another. In incomplete metamorphosis the immature stages resemble adults in structure and are termed nymphs. In complete metamorphosis, the immature stages (larvae and pupa) do not resemble the adults.
Metazoa
A subkingdom of animals consisting of all multicellular animals, in which cells are differentiated to form tissue. The subkingdom includes all animals except Protozoa.
Microfilaria
The juvenile, first-stage larva of filarial nematodes.
Microgametocyte
The cell that gives rise to microgametes.
Miracidium
The ciliated larva that emerges from the egg of digenetic trematodes.
Molecular mimicry
The acquisition by a parasite of a body surface that manifests host-like molecules.

Molt (=ecdysis)
Shedding of an external covering (e.g., integument, cuticle or exoskeleton), and the replacement of a new one that accommodates for the growth and development of the pre-adult stages.

Monoecious
Both male and female sex organs in one individual (i.e., hermaphroditic).

Monoxenous
Having a single host in the life cycle.

Monozoic
Refers to tapeworms that possess only a single set of reproductive organs.

Multiple parasitism
Infection or infestation of a host by more than one species of parasite.

Multivoltine
Having a number of generations in a year.

Mutualism
A symbiotic relationship in which both symbiotic individuals benefit from each other.

Myiasis
A condition caused by infestation of the body with fly larvae.

Nagana
A disease of domestic ruminants, caused by *Trypanosoma brucei brucei*, *T. congolense* and *T. vivax*.

Naiad
The aquatic, pre-adult stage of an insect that has incomplete metamorphosis.

Natural immunity
The type of immunity conferred by the presence in an organism of certain naturally occurring antibodies to specific antigens.

Nymph
The terrestrial, pre-adult stage in the life cycle of an insect with incomplete metamorphosis.

Obligate parasite
A parasite that requires a host for the completion of its life cycle.

Oncosphere
The motile, six-hooked embryo (hexacanth) of a tapeworm that is contained in the egg membranes.

Oocyst
A stage in the life cycle of certain members of the phylum Apicomplexa in which the zygote secretes a resistant covering around itself. The encysted form of the ookinete which occurs on the stomach wall of *Anopheles* spp. mosquitoes infected with malaria.

Ookinete
The zygote in the life cycle of certain members of the phylum Apicomplexa following the fusion (syngamy) of macro- and microgametes. The term ookinete most often refers to the motile stage of *Plasmodium*, which is seen in the midgut of the mosquito shortly after syngamy.

Ootype
A specialized region of the flatworm oviduct that is surrounded by Mehlis' gland.

Open circulatory system
The system in which blood flows slowly through large sinuses (hemocoel) back to a dorsal, tubular heart.

Operculum
A lid-like structure at one end of the eggshell of many digenetic trematodes and some cestodes.

Oriental sore
A disease, also known as cutaneous Leishmaniasis, caused by *Leishmania tropica*.

Oviparous
Producing eggs that hatch after leaving the body of the mother.

Ovoviviparous
Producing eggs with persistent membranes through which the young escape while still within the body of the mother.

Ovum (pl. ova)
The female germ cell.

Parabasal body
The Golgi complex of the protozoan order Trichomonadida.

Parasitemia
The presence of parasites in the blood (e.g., malaria schizonts in red blood cells).

Parasitism
An association between two species in which the smaller (parasite) is physiologically dependent on the larger (host). The host may be adversely affected by the parasite.

Parasitophorous vacuole
A clear space between an intracellular parasite and the host cell cytoplasm.

Paratenic host (=transport host)
A host in which a parasite resides but does not develop and which is not physiologically essential for the completion of the life cycle.

Parenchyma
In flatworms, the mesodermal tissue filling all available body spaces.

Paroxysm
The fever-chills syndrome in malaria occurs cyclically every 36 to 72 hours depending on the species of *Plasmodium*. The onset of fever corresponds to the release of merozoites and toxic materials from infected red blood cells. The shaking chills are manifested during schizont development.

Parthenogenesis
Development of an organism from an unfertilized egg; common in insects such as aphids and in some nematodes such as *Strongyloides*.

Parthenogenic
Capable of unisexual reproduction; (i.e., without fertilization).

Pathogenic
Production of tissue changes or disease.

Pathogenicity
The ability to produce pathogenic changes.

Periodicity
Recurring at regular time periods.

Peritoneum
A thin membrane of mesodermal origin that lines the body cavity of vertebrates (and some other higher metazoans) and supports the viscera or organs of the body cavity.

Peritrophic membrane
A covering that forms around the blood meal of a hematophagous insect; digestion usually takes place within the membrane; the membrane sometimes serves as a barrier to a parasitic agent moving from the blood to the tissues of the host.

Phasmid
Sensory pit located on the posterior part of nematodes of the class Secernentea.

Phoresis
A form of commensalism in which one organism is mechanically transported by another. The relationship is nonobligatory.

Plasma cell
An effector B cell that secretes into the circulation antibodies of the same specificity as its cell surface receptors.

Plerocercoid
A larva or metacestode developing from a procercoid in the life cycle of pseudophyllidean tapeworms. This larval stage, which is infective to humans, develops in fresh-water fish.

Polar filament
A coiled filament or tubule in the spore of certain protozoa (e.g., Myxozoa and Microspora). The filament is extruded when the spore is ingested by a host.

Polyembryony
A process in which a zygote gives rise to more than one embryo.

Polyzoic
Consisting of more than one zooid or animal or proglottid.

Procercoid
The larval stage that develops from the coracidium of the pseudophyllidean tapeworm *Diphyllobothrium latum*. The procercoid is found in the body of a freshwater crustacean that serves as the first intermediate host.

Proglottid
One of the segments of a tapeworm strobila containing male and female reproductive organs when mature.

Promastigote
A morphological form of flagellates in which the kinetoplast is located at the anterior end of the organism, and there is no undulating membrane.

Prophylaxis
Procedures that are carried out to prevent the transmission of disease.

Protandry
The condition in which the male gonad matures before the female gonad, or a single gonad first produces sperm and then ova.

Protoscolex (protoscolices)
The immature scolex found in coenurus and hydatid larvae of tapeworms.

Pruritus
Intense itching.

Pseudocoelom
A body cavity of a metazoan that is b;t completely lined with mesoderm.

Pseudocyst
A cluster of am2'tigotes of *Trypanosoma cruzi* in a muscle fiber. A cyst-like structure filled with bradyzoites of *Toxoplasma gondii*.

Pseudopod
A protoplasmic extension of amebae which allows them to move and engulf food.

Pupa (pl. pupae)
The encased resting stage (e.g., cocoon) between the larva and adult (imago) stages of certain insects.

Quarantine
Limitation in the freedom of movement of humans or animals in order to contain the spread of a disease.

Radula
Rasping structure inside the mouth of mollusks of the classes Gastropoda, Cephalopoda, Amphineura, and Scaphopoda; the pattern of teeth on the radula is characteristic of a species.

Rectal prolapse
Weakening of the rectal musculature resulting in a "falling down" of the rectum; occasionally seen in heavy whipworm infections, particularly in children.

Redia (pl. rediae)
A larval form of digenetic flukes that arises asexually from within a sporocyst or a primary redia.

Relapse
The recurrence of symptoms of a disease after an abatement of weeks or months.

Reservoir host
An animal that harbors a species of parasite from which humans may become infected.

Resistance
The ability of an organism to withstand infection or the successful establishment of a parasite.

Retrofection
A process of infection in which a parasite leaves the body of the host and then returns almost immediately, frequently by penetrating the skin (e.g. the nematode Strongyloides).

Rhabditiform larva
Non-infective, feeding, first-stage larva of some nematodes.

Romano's sign
Early symptoms of Chagas' disease, consisting of unilateral, periorbital swelling (edema) and conjunctivitis.

Rostellum
The rounded protuberance on the apex of the scolex of some tapeworms that usually bears a circular row hooks.

Sarcodina
A subphylum of Protozoa containing amebae that possess pseudopodia for movement.

Schiffner's dots
Fine granules distributed throughout erythrocytes infected with *Plasmodium vivax*.

Schizogony
A form of asexual reproduction characterized by multiple nuclear divisions followed by cytoplasmic divisions and the resulting formation of a large number of daughter cells.

Schizont
A multinucleated cell undergoing schizogony prior to cytoplasmic division.

Scolex (pl. scolices)
Anterior end of a tapeworm that attaches to the host intestinal wall. The scolex may be comprised of suckers and hooks.

Scrub typhus
Rickettsial disease transmitted by chigger mites.

Scutum
A chitinous shield or plate covering the dorsal surface of hard ticks.

Serology
The study of antibody-antigen reactions in vitro, using host serum for study.

Serum
The fluid part of vertebrate blood after the fibrin has been removed.

Shell (yolk) glands
Clusters of cells that synthesize substances essential to eggshell formation.

Sleeping sickness
A disease caused by *Trypanosoma rhodesiense* or *T. gambiense* in Africa, or any one of several arboviruses transmitted by mosquitoes and causing encephalitis.

Sparganosis (or sparganum)
Infection with a plerocercoid or larval stage of a tapeworm belonging to the order Pseudophyllidea.

Spicule (=copulatory spicule)
A sclerotized structure of male nematodes used during copulation.

Spore
A resistant stage that is formed internally by the mother cell.

Sporoblasts
Cells that divide into sporozoites.

Sporocyst
The larval stage of digenetic trematodes developing in the first intermediate host.

Sporogony
Sexual reproduction of Apicomplexa. Production of spores and sporozoites.

Sporozoite
The infective or transfer stage of *Plasmodium* and other members of the phylum Apicomplexa

Stenoxenous
Having a narrow host range.

Strobila
In tapeworms, a chain of proglottids or segments formed by budding.

Superparasitism
The infection or infestation of a host by more individuals of a single species of parasite than the host can support.

Swimmer's itch (=cercarial dermatitis)
A skin hypersensitivity reaction in response to the penetration or attempted penetration by non-human schistosome cercariae.

Sylvatic
Refers to an animal (or disease) cycle that exists in the wild.

Syngamy
The union of gametes.

Symbiosis
The intimate association of two different species of organisms exhibiting metabolic dependence by their relationship.

Syzygy
End-to-end joining of two or more gamonts of members of the subclasses Gregarina and Coccidia.

T cell
A specialized lymphocyte, processed through the thymus, that elicits cell-mediated reactions.

Tachyzoite
A form of merozoite in Toxoplasma, found in parasitophorous vacuoles of vertebrate hosts. Rapidly growing meront or zoites characteristic of the early stage of infection with Toxoplasma and related organisms of the phylum Apicomplexa. Rapidly growing intracellular trophozoites of *Toxoplasma gondii*.

Temporary host
A host on which an arthropod (adult or larval form) reside temporarily in order to feed on blood or tissue.

Temporary parasite
A parasite that visits a host at intervals and only for relatively short period; examples are mosquitoes and ticks.

Transport host
See paratenic host.

Trophozoite
The motile, growing, and feeding (vegetative) stage of a protozoan stage of a protozoan that maintains the population within the host.

Trypomastigote
A hemoflagellate form with a kinetoplast located posterior to the nucleus and an elongated undulating membrane extending along the entire body. This form is seen in the blood of humans with trypanosomiasis and as the infective stage in the insect vectors.

Tularemia
A bacterial disease of humans transmitted by tabanid flies and for which rabbits often serve as sylvatic reservoir hosts.

Undulating membrane
That portion of a plasma membrane or cytoplasm of a flagellate that is drawn away from the cell like a fin along the outer edge of the. The membrane moves in a wave-like pattern.

Uniramous
Having one branch or ramous.

Univoltine
Having one generation per year.
Vagina
An organ that joins the oviduct and carries sperm from the genital atrium to the oviduct.
Vector
Any organism that actively transmits a disease-producing organism from an infected to a non-infected individual. A mechanical vector is one in which the parasite neither multiplies nor develops (passive transmission). A biological vector is one in which the parasite either multiplies or develops;
Vermicle
The infective stage, analogous to sporozoite, of Babesia spp. from ticks.
Vertical transmission
Transmission of a parasite from one generation to the next through the egg or in utero.
Virulence
The ability of a parasite to produce pathogenic effects or to invade the host and become established.
Viscera
Any of the large organs in the body cavities of vertebrates.
Visceral larva migrans
Migration of second-stage larvae of nematodes in the internal organs of unnatural hosts.
Vitellaria
Clusters of cells that synthesize substances essential to eggshell formation.
Winterbottom's sign
Symptom of African sleeping sickness characterized by enlarged, sensitive cervical lymph nodes.
Yaws
A fly-transmitted spirochete disease.
Yellow fever
A viral disease transmitted by the mosquito *Aedes aegypti*.
Zoonosis (pl. zoonoses)
A disease of animals that can be transmitted to humans.
Zoonotic agent
An organism that causes a zoonosis.

Index

Embryonation 41, 58, 73, 126
Embryophore 56, 57, 124, 126
Endemic 12, 14, 47, 58, 87, 96, 98,
      102, 107, 113, 126
Eosinophilia 3, 86, 87, 88, 126
Epidemic 102, 112, 126
Epidemiology 10, 126
Epimastigote 7, 13, 126
Epizootic 102, 127
Espundia 10, 17
Excystation 18, 20, 23, 24, 41, 45, 47,
      127
Exoskeleton 5, 100, 122, 127, 132

F

Flagella 5, 7, 18, 21, 22, 127
Flagellum 7, 122, 123, 124, 127

G

Gamete 6
Gametogony 29, 122, 127
Gamogony 25, 26, 28, 127, 131
Genital atrium 37, 127, 138
Genital pore 37, 71, 127
Gravid 55, 58, 60, 61, 68, 69, 84, 91,
      105, 118, 121, 122, 128
Ground itch 80, 128

H

Helminth 47, 126, 128
Hematophagous 128, 134
Hematuria 102, 128
Hemocoel 33, 69, 89, 91, 93, 128,
      133
Hemozoin 33, 128
Hermaphroditic 37, 128, 132
Heterogonic 78, 128
Heteroxenous 25, 72, 128
Hexacanth embryo 123, 128
Homogonic 77, 128
Host 1-4, 6, 7, 9-16, 18, 20, 25-28,
      31, 34, 36-38, 40, 41, 43-45, 47,
      49, 51, 55-58, 60, 62-65, 69, 70,
      72-74, 76-81, 84, 86, 89-91, 93,
      96-99, 105, 107, 109-113, 115,
      117-138
Host specificity 2, 128

Hydatid 55, 58, 63-68, 129, 135
Hydatid sand 65, 129
Hyperplasia 3, 17, 43, 46, 94, 129
Hypersensitivity 121, 125, 129, 137
Hypertrophy 3, 95, 129

I

Imago 128, 129, 135
Immunity 2, 28, 29, 86, 121, 123,
      129, 132
Immunopathology 28, 129
Incidence of infection 58
Incidental host 121
Incubation period 16, 17, 34, 130
Indirect life cycle 56, 130
Infection 1, 3, 4, 9, 10, 12, 14-18,
      20-22, 24-26, 28, 29, 33-35, 41,
      44-48, 51, 54, 57-61, 63, 65, 69,
      74-76, 78-88, 91, 93, 94, 96-98,
      106, 121, 125, 126, 129, 130,
      132, 135-137
Infective stage 3, 25, 71, 86, 91, 98,
      130, 138
Infestation 1, 111, 113, 115, 130,
      132, 137
Inflammation 3, 9, 12, 17, 41, 44, 54,
      82, 84, 88, 99, 120, 125,
      126,130
Intermediate host 2, 3, 27, 31, 37, 38,
      40, 43, 47, 49, 51, 56-58, 60,
      63, 65, 70, 89, 90, 97, 113, 130,
      134, 137

K

Kala-azar 10, 14, 16, 17, 102, 130
Kinetoplast 7, 126, 130, 135, 138

L

Larva 37, 56, 57, 60, 81, 82, 85, 87-
      90, 92, 103, 104, 118, 119, 123,
      125-128, 130-132, 134-136, 138
Leishmaniasis 15-17, 102, 115, 117,
      127, 130, 131, 133
Lumen 23, 28, 69, 74, 77, 84, 89, 94,
      100, 131

Index

Index